# THE BRIDGE

Everyone's favorite, of course. This is the story of
The Golden Gate Bridge — from first dreams and impossible
challenges, to early political and legal struggles, formation of
the District, financial growing pains, the inspired design
and tenacious construction during the Great Depression,
right up to the glorious Opening Day, May 27, 1937 . . .
and that was the easy part.

This is also the story of what's happened since. Here is
how the world's most beautiful engineering triumph grew
into a necessary, working reality — a full-service
regional transportation system.

As you will discover when you visit the masterpiece in
person or on these pages, there is much more to the Bridge
than just, "The Bridge." This is the story of a structure
that bridged more than a channel of water. It has also
spanned six generations, and several eras. Today, more
than ever, it mirrors the rugged and changing
American historical landscape.

THE BRIDGE celebrates this continuing achievement.

# THE BRIDGE: A Celebration

# CONTENTS/TRIVIA

| Question: | Page: |
|---|---|

# THE BRIDGE: A Celebration

## The Collector's Anniversary Edition
### First Edition

THE BRIDGE: A Celebration
On the Occasion of the Sixtieth Anniversary of the Opening of
THE GOLDEN GATE BRIDGE

Published by James W. Schock
Copyright © 1997 by James W. Schock

JAMES W. SCHOCK
SEVENTY-TWO LOCUST AVENUE
MILL VALLEY, CALIFORNIA 94941-2131

ISBN: 0-9660989-0-0

THE BRIDGE: A CELEBRATION
is available at special discounts for bulk purchases
for sales promotions, premiums, fund-raising, or
educational use. For details, contact:

CORPORATE SALES DIRECTOR
Seventy-Two Locust Avenue
Mill Valley, CA 94941-2131
Fax: 415.383.5685 (USA)

James W. Schock
Presents

# THE BRIDGE:
## A Celebration

## The Golden Gate Bridge at Sixty

Produced by
James W. Schock

Golden Gate International, Ltd.
Mill Valley, California, USA

# What They Said

## Before the Bridge Was Built

"To this Gate, I gave the name *Chrysopylae*, or Golden Gate, for the same reasons that the harbor of Byzantium was named *Chrysolceras*, or Golden Horn."

John C. Fremont

"Building a bridge is a war with the forces of Nature."

Joseph Strauss

"Federal experts believe it will be impossible to put piers at this point owing to strong currents and great depths."

*San Francisco Chronicle*
Quoting sources, 1920

(The Bridge will) "open the gateway to thousands of home seekers now held back by lack of proper transportation facilities."

Mayor of San Rafael, 1922

"Everybody says it can't be done and that it would cost over $100,000,000 if it could be done."

M. M. O'Shaughnessy
San Francisco City Engineer

"In the age of radio and aeroplane, the ferry is too slow and uncertain a method of transportation."

Charles Derleth
U-C Berkeley Professor

"It is possible to bridge the bay at various points, but at only one can such an enterprise be of universal advantage -- at the water gap, the Golden Gate -- giving a continuous, dry-shod passage around the entire circuit of our inland sea."

James Wilkins

"It would be two and a half times larger than any similar bridge in the world. The towers would be ten feet higher than the Eiffel Tower. There is no better place for the eighth wonder of the world than Northern California."

Joseph Strauss
Speech, March 1933

"San Francisco is one of the few cities that has all the energy, all the wealth, all the courage and all the ability that is needed to undertake and carry this project to success."

Joseph Strauss
May 16, 1924

"We'll take the bonds. We need the bridge."

A. P. Giannini
Bank of America, 1932

A. P. Giannini: "How long will the bridge last?"
Joseph Strauss: "Forever."

Conversation, 1932

(It has been) "a long and tortuous march . . . the worn sandals of weary crusaders sink to rest at the foot of a long-receding goal. It took two decades and two hundred million words to convince people the bridge was feasible."

Joseph Strauss
February 1933

# ACKNOWLEDGEMENTS

*E*very book is a collaboration. This one certainly was. Our unique editorial approach required close alliance with and assistance from the busy, helpful people at the Golden Gate Bridge, Highway and Transportation District.

Most notably of these, Linda Mitchell, Director of Marketing, became involved early-on with the project. Her encouragement, guidance and cooperation were vital to the production of this work and our thanks do not begin to measure her contributions. Public Information Officer, Mary Curie's suggestions and help in tracking down key details must be included. A constant sense of humor and sharp eyes were supplied by Timothy J. Moore. In the Graphics Department, Arline Laesa and Bob David helped sort through the chaos of finding and reproducing photos.

When this book was being planned, I mentioned the project to Herb Caen, famed (and now fabled) columnist of the *San Francisco Chronicle*, whom I met about the same time I first crossed the Bridge. A light came to his eyes and he said, "Beautiful ladies make beautiful books. Be good to her." Thanks, Herb; miss you.

Janet Carter of Bank of America provided gracious and prompt research and photos. Laura Hauck, of the law firm of Hanson, Bridgett, Marcus, Vlahos & Rudy was of immense help dotting the i's and crossing the t's. Michael Slaughter and Bob Wilson made part of the journey from idea to finished product.

Publishing the book would not have been possible without the insight and guidance of David Poindexter. Others to whom we wish to express our gratitude include Gordeen Davidson, Ed Addeo, Trubee, Jeffrey, David, and Will Schock. Thanks, too, to Al and Doreen Burgin, publishers of *The Commuter Times*, for loosening some deadlines in order that work could progress on this book.

**Photo Credits.** The photos appearing in this book are from the archives of the Golden Gate Bridge, Highway and Transportation District and are owned by them. The A.P. Giannini photo courtesy Bank of America. **Cover Photo by Jerry Littlejohn.** Printed in the USA by American Litho, Hayward, California.

# DEDICATION

## THE BRIDGE: A CELEBRATION

. . . is respectfully dedicated to all of
those who first dreamed about,
fought for, designed and who finally
built the Golden Gate Bridge; to
everyone who helps operate and
maintain it and its associated transportation
endeavors; and all of us to whom
the Bridge is something very special.

# Mission Statement

The District's mission is to provide safe, efficient, reliable means for the movement of people, goods and services within the Golden Gate Corridor.

*This photo was taken within a few weeks of the first time I saw the Golden Gate Bridge. We're both older now.*

It was one of those dark, gray afternoons.

The first time I saw the Golden Gate Bridge, she was a beautiful teenager. She came into view, graceful, inviting, a flame against the fading light of a December afternoon. Then and there, I lost my wildly beating heart and I have been in love with her ever since.

Now sixty years old, she retains the old magic. In the midst of a world that is rapidly changing around her, she seems eternal, peaceful in her beauty, assured of her accomplishments, and confident of her future, of which more in a moment.

Structurally, the Golden Gate Bridge is as good now as the day she opened, May 27, 1937. This is a lasting tribute to her designers and builders, and to the legion of faithful who continue to watch over her. Mind, she has had her share of repairs. Her entire roadway deck was replaced. Her suspender cables were replaced. Every rivet was removed and a new bolt installed. Today, she is lighter stronger, and better-lighted than Opening Day. On her Fiftieth Anniversary, so many people crowded onto the roadway, the deck flattened. Lady-like, she issued no complaint, sustained no harm.

This year about ten million people will visit the Bridge. For the past few years, more than 40 million vehicles annually have crossed her span, but the Golden Gate Bridge has no time to pause and celebrate — so we're doing that for her. The purpose of this book is the revel in her past, enjoy her present and look ahead to her breathtaking future.

## A New Kind of Book

When we began the task of finding a new and special way to bring meaning to this significant anniversary, we wanted to detail the never-ending work of the the men and women who manage, operate, and maintain the Bridge, and the associated transit business which has bloomed as a natural outgrowth of the graceful Bridge that spans the Golden Gate.

Rather than research old books, re-write old news stories, rework old interviews, we went to the Source — the documents of the Bridge itself. We have allowed the Bridge tell her own story, in her own words. We discovered that in the simplicity of this concept, there was great strength and, for you readers, there is meaningful reward.

**Inasmuch as humanly possible, we have let the Golden Gate Bridge story be told by the founders, builders, managers, and employees who did (and still do) the work. The Dreamers, the Builders, the Caretakers, the Planners.**

In early pamphlets and booklets, we found original text written by the people who were there at the time, actively involved in the endless process that is the Golden Gate Bridge, Highway and Transportation District. In the Annual Reports of the District, we unearthed the genuine story, one in which the problems of the day were solved by the people of the day.

## Authenticity The Key

While the use of language, punctuation and capitalization evolve from year to year, these documents echo the pride, satisfaction, frustration, and gritty hard work of people who had to solve not only the political, financial, and societal problems each year, but who also had the demanding challenge of sustaining the Dream, protecting the Bridge — the Lady — from physical harm and political neglect.

During many of the years, the job of these Keepers of the Flame, was not a pleasant or even popular one. Despite the hue and cry of special interests, they stayed the course, kept the District fiscally healthy and the Bridge physically sound. This has been a hallmark of the Bridge — from outnumbered but determined men who believed a span could be built across the dangerous entrance to San Francisco Bay, to the gifted designers and engineers of the early days, to the Board, Staff, and Workers of today who face the challenges of Time and explosive growth which daily jams far more vehicles onto the Bridge than was planned or ever imagined.

This book is a "Thank You" to the Bridge Family for protecting a structure that is more than just a bridge to all of us. Our salute is more than an homage to the most beloved engineering achievement of the modern world. It is a celebration in which we wish the The Bridge and its People a "well done" for Sixty Glorious Years, and for their compassion, skill, and dedication that assures us it will endure for decades to come.

James W. Schock
Mill Valley, California
Spring 1997

# The Bridge Writes Its History

*Early in its history (probably 1940), the Bridge published a small pamphlet describing what the District regarded as some of the important details of its early beginnings. Written by W. W. Felt, Jr., the first Secretary of the District, and reprinted here for the first time, the story is interesting in its choice of detail, its richness of language and its gentle spirit. Unlike other District documents, this one begins with a legendary tale. The use of the terms, "Indian" and "Red men" were standard at the time, having since been replaced in modern lexicon by "Native American." No racial slight is intended nor implied, as is evident by the considerate tone of the story.*

# ANCIENT LEGEND

ccording to an ancient Indian legend, California's great interior valleys were at one time an immense lake, while what is now known as San Francisco bay was a fertile valley – the happy hunting ground of the Red men.

This valley was separated from lake and ocean by ranges of tall mountains. Long before the white man came to these shores, legend has it that the Sun God was a personage among the Indian inhabitants, and frequently, in his flight from east to west, paused to enjoy a visit with his aboriginal friends.

On one such occasion, the deity, falling madly in love with the beautiful princess of the tribe attempted her abduction, and taking a hurried departure, is purported to have stubbed his toe on Mt. Diablo. Plunging headlong, his captive was killed. Deeply remorseful, he tenderly laid the body atop Mt. Tamalpais, and strangely enough, to this day the contour of the mountain top vaguely outlines the reclining form of a woman.

In his headlong plunge, the Sun God's toe clove an opening between the lake and valley, while his arm tore a deep gash in the mountain range separating the valley from the ocean.

A way was now made for inundation, and the waters pouring in from lake and oceans, formed what is now the Golden Gate strait and the magnificent harbor of San Francisco bay.

*From a brochure written by Golden Gate Bridge and Highway District Secretary, W. W. Felt, Jr., circa January 1940.*

# BY WHATEVER MANNER FORMED

*I*gnoring the Indian legend, the geologists hold that the Golden Gate was at one time a river's gorge, through which flowed the greater part of California's drainage as a fresh water stream.

They also affirm that the mouth of the river discharged over a delta outside the gorge which extended as far out as the Farallon Islands, and that this delta was molded by the action of wind and wave into a vast embankment extending many miles along the coast.

A gradual subsidence of the coast line in the region of the outlet of the river permitted a slow invasion of ocean water into the river gorge, which eventually flooded the valley and formed the bay.

The geologists conclude the degree of subsidence to be about 350 feet, drawing this deduction from the fact that the greatest depth of the Golden Gate is 378 feet, and that, when the gorge functioned as a river channel, its bottom was probably not more than 30 feet below sea level.

Be that as it may – and by whatever manner formed – the barrier thus created remained a challenge to man for ages to come. Crossing from shore to shore was not only hazardous, but entailed considerable effort as well.

*From a brochure written by Golden Gate Bridge and
Highway District Secretary, W. W. Felt, Jr., circa January 1940.*

# DEMAND FOR RELIEF

As time went on, a steadily growing population came to demand speedier and safer transportation over this stretch of water; before this demand, row boat and sail gave way to the more modern steam ferry boat, which finally became the accepted medium, serving the traveler for a long period of time. With ever increasing automobile travel, however, traffic congestion was vastly aggravated, and although the ferry system was improved and augmented, the service remained expensive and inadequate for the modern need.

The demand for relief from this condition finally crystallized in the call for a public meeting which took place in the City of Santa Rosa in January 1923. At this meeting legislation was determined upon, intended to set the wheels of legal machinery into motion which would have for its objective the building of a bridge over the Golden Gate. An enabling act was drafted, introduced into the 1923 California legislature, was finally passed and was approved by Governor Richardson on May 25th of the same year.

Upon petition of electors, proceeding under the provision of law applicable to passage of ordinances by the initiative, the Boards of Supervisors of Marin, Sonoma, Napa, Mendocino and Del Norte Counties enacted the necessary ordinances creating the Golden Gate Bridge and Highway District.

*From a brochure written by Golden Gate Bridge and
Highway District Secretary, W. W. Felt, Jr., circa January 1940.*

# THE OPPRESSIVE YOKE

Opponents of the Bridge, encouraged by certain vested interests, immediately brought court actions, calculated to destroy or at least delay for as long as possible this popular movement of the people to free themselves from the oppressive yoke of high ferry tolls and intolerable traffic conditions. After decisions had been rendered in these actions, favorable to the District, and under the date of December 4th, 1928, the Secretary of State issued his Certificate of Incorporation of the District, including each of the aforesaid counties, with certain lands excluded in the Counties of Mendocino and Napa.

The Boards of Supervisors of the Counties of the District then appointed Bridge Directors from their respective counties. The Directors thus appointed met in City Hall in San Francisco on January 23, 1929, and organized . . . and officers of the Board were elected. Later in 1930, the membership of the Board was increased to fourteen and two members were appointed from San Francisco.

After organizing as aforesaid, serious consideration was given to preparation of plans and the necessary financing, and the statistical hereafter presented, records the steps taking in their accomplishment.

For preliminary financing, on July 24, 1929, a three-cent tax levy was made on taxable property in the District, estimated to raise $240,000.

August 15, 1929, The Chief Engineer was appointed and an Engineering Board created, comprising Joseph B. Strauss as President, and Consulting Engineers Leon S. Moisseiff, O.H. Ammann and Charles Derleth, Jr. The Engineering Board met in San Francisco almost immediately following appointment.

On November 15, 1929, a contract was let to Longyear Exploration Co., for $29,940.00 for 1300 lineal feet of casing and 800 lineal feet of 2-inch diamond drill rock cores, at site of San Francisco pier.

*From a brochure written by Golden Gate Bridge and Highway District Secretary, W. W. Felt, Jr., circa January 1940.*

On April 29th, 1930, plans for location of piers, anchorages and clearances were presented to the Directors, and by them submitted to the Chief of Engineers of the U.S. Engineer Corps and the Secretary of War for their approval, and it was also ordered that application be made to the Secretary of War for extension of roads across the Presidio of San Francisco and the Fort Baker military reservation in Marin County, to serve as approach roads to the Bridge.

A second tax levy of two cents was made on July 9th, 1930, estimated to raise an additional $160,000.00 necessary for preliminary financing.

On August 27th, 1930, the Directors approved plans revised by the Engineers to meet requirements of the War Department, and on the same day the Chief Engineer filed his report with the Directors, which contained his statement of construction cost of the Bridge, including administrative costs, interest during construction, and for the first six months of operation as $32,815,000.00. The Directors further, on this day, adopted resolutions, fixing details for submission to the electors of the District, at the General Election, to be held November 4, 1930, of a proposal for issuance of $35,000,000 bonds of the District, estimated as sufficient for financing the construction of the Bridge.

The Chief of Engineers of the U.S. Engineer Corps and the Secretary of War, on September 8th, 1930, approved the revised plans submitted to and approved by the Directors on August 27th, and under the date of October 27th, 1930, the Secretary of War issued his permit for construction of the approach roads.

On November 12, 1930, the Board of Directors met in the City Hall at San Francisco as a canvassing Board and canvassed the returns of the vote cast at the election held November 4, 1930, on the proposal for the issuance of $35,000,000 of District bonds. The Directors then adjourned to November 19th, 1930, when they met again in the City Hall and canvassed the absentee votes, and declared the result of the District bonds as 145,657 "yes" and 46,954 "no".

*From a brochure written by Golden Gate Bridge and Highway District Secretary, W. W. Felt, Jr., circa January 1940.*

# $35,000,000

*D*uring the next several months, detailed plans and forms of contract were prepared, and preparations made for calling for bids on the various units of work. Earnest efforts were made to find a buyer for the bonds. On March 12, 1931, a contract was let in the sum of $52,800 for an additional 1000 lineal feet of 2" core diamond drill boring at the San Francisco pier site.

On April 22, 1931, the Directors ordered advertisements be made calling for bids on the ten units into which the work had been subdivided, to be received and opened on June 17, 1931. The bids were received and opened as aforesaid and the total amount of the lowest bids received on each of the units was $24,455,299.93. After adding to the total of the bids certain minor miscellaneous items which could be closely estimated by the Engineers, it was easily seen that the $35,000,000 bond issue was ample to finance construction of the Bridge.

Also on June 17th, following the receipt and opening of bids as aforesaid, the Directors ordered that $6,000,000 par value, 4 3/4%, Series A District bonds be offered for sale, bids to be received and opened on July 8th, 1931. One bid only, that of Dean Witter & Co., was received and taken under advisement.

The Directors at a meeting held on July 16th, 1931, rejected this bid on the ground that it was not a direct offer to purchase said bonds unconditionally.

At the same meeting, bids on the various units of the work, received on June 17th, 1931, were accepted, conditioned that the District had an option of entering into contracts with the various contractors within six months after July 16th, 1931, and further conditioned that during this period not less that $6,000,000 par value District bonds could be sold. At this meeting the Bank of America offer to purchase the $6,000,000 par value of bonds, for par, accrued interest to date of delivery, plus a premium of $7,500.

The Secretary of the District, following receipt of the Bank of America to purchase $6,000,000 par value of bonds, advised the Directors that on account of conditions imposed and contained in said offer, that he would refuse to sign any of said bonds, if said offer were accepted.

*From a brochure written by Golden Gate Bridge and Highway District Secretary, W. W. Felt, Jr., circa January 1940.*

*[Note: Remember, the Secretary of the District is the person who is writing this. What he neglects to mention is his refusal to sign the bonds was orchestrated at a meeting on July 16th by the District itself. The BofA bid had only one condition — a court test that, in effect, would verify the District's right to levy taxes as collateral for the bonds. This and other suits that were filed troubled the Directors, not because of their alleged merit, but because the District was perilously close to bankruptcy at the time. With the clock running on expenses of nearly $4,500 per month, the cash from the BofA bonds was their only source of income. Secretary Fell's story continues with all the intrigue and suspense of a modern mystery.*

A resolution was then adopted by the Directors, ordering the Attorney of the District to commence an action to compel the Secretary to sign such bonds. Such action was commenced in the Supreme Court of California by Petition for Writ of Mandate, to compel the Secretary to sign the bonds of the District. "Golden Gate Bridge and Highway District v. Felt, 82 Cal. Dec 683". The matter was strongly contested before the Court and finally resulted in the Court making its order directing that the Writ of Mandate prayed for, issue, directing that the Secretary sign the bonds.

Following this, taxpayers' suits were brought in the United States District Court . . . in which actions raised by (cited) taxpayers under the Constitution of the United States, to the formation of the District, the proceeding for the issuance of the bonds, and the right to the District to levy and collect taxes.

On July 16th, 1932, (a judge) rendered judgment in said actions holding against the plaintiffs. After rendition of judgement, an aroused public and press demanded that no further actions be taken in Court to delay beginning of the work, particularly in view of the serious unemployment conditions. The opposition yielded to this pressure, and abandoned any further attempt for delay through the Courts. Thus ended long years of litigation, in which every possible question that could be raised in opposition to the District, was presented in the courts, and decided favorably for the District.

*From a brochure written by Golden Gate Bridge and Highway District Secretary, W. W. Felt, Jr., circa January 1940.*

# CONTRACTS WERE LET...

So much time having elapsed since receiving bids in June 1931, all bidders were released from their bids with the exception of Contract I-A, for furnishing and erecting the steel for the towers and main span. New bids were called for to be received and opened on October 14, 1932, excepting for cement, which was to be included in the contractor's proposal for concrete; also omitted were the toll terminal and final painting.

The bids when opened, totaled, including the bid on Contract I-A, $23,843,905. On November 4th, 1932, the Bank of America agreed to purchase $3,000,000 par value of Series "A" 4 3/4% Bonds, and on the same day contracts were awarded on the basis of the bids received and opened on October 14th. The same Bank took another $3,000,000 par value of the same series of bonds on November 16th, 1932, at an average price of $96.233 1/2, being the same prices as was paid for bonds purchased on November 4th.

On the basis of the bids received on October 14th, contracts were let for work excepting the Sausalito Approach work. This road was later built by the District with W.P.A. assistance. January 5, 1933, marked the actual commencement of construction.

April 26th, 1933, an agreement was entered into between the District and a syndicate of Investment bankers, composed of Bankamerica Company, Dean, Witter & Company, Blyth & Company, and Weeden & Company, for the marketing of the remaining $29,000,000.

*[We pause here in Secretary Felt's review of the formation of the District and its early problems. First, for some additional details not included in his report. Second, for the graphic story of the construction of the Golden Gate Bridge. We will rejoin Mr. Felt's story in-progress after the completion of the Bridge.]*

## SOME OTHER DETAILS

*M*any believe it is possible to trace the distant beginnings of the Golden Gate Bridge to the middle of the 19th century — and tie it directly to the Gold Rush of 1849, which saw the village of Yerba Buena with a population of about 400 become the city of San Francisco and, by year's end, home to more than 35,000 fortune-seekers.

With prosperity came the realization that the growth of the San Francisco peninsula would require commerce from the rich lands to the east and north. Oddly, the idea of a bridge across the Golden Gate came from one of the city's early characters, an unlucky Gold Rush merchant whose bankruptcy pushed him over the edge to lunacy. He declared himself "Norton I, Emperor of the United States and Protector of Mexico."

He printed his own money which was accepted by restaurateurs of the day in good humor, allowing Emperor Norton to "rule" in modest prosperity while both he and the city's merchants avoided the mutual embarrassment of the handout.

In 1869, an announcement appeared in the Oakland Daily News under Norton's imprimatur ordering a suspension bridge to be built from San Francisco to Yerba Buena Island (an anchorage of the present-day Bay Bridge to which is attached the man-made Treasure Island) and "from thence to the mountain range of Sausalito." He also wanted his bridge to continue out to the Farallones and "to be of sufficient strength and size for a railroad."

Nobody paid much attention.

However, about three years later, a heavyweight financier reported to the Marin County Board of Supervisors that his engineers had prepared drawings and estimates for a suspension bridge across the Golden Gate. The man, Charles Crocker, was the railroad tycoon who had built his Central Pacific Railroad from Sacramento, over the Sierra to Utah. Crocker, one of the "Big Four" (the others being Leland Stanford, Mark Hopkins and James Flood), discarded the idea later in favor of putting his trains on huge steam barges.

# THE COMMUTER & THE ENGINEER

*I*t was more than forty years later when the idea once again came forward, this time from a newspaperman who worked on the San Francisco Bulletin. James Wilkins, a San Rafael resident, commuted to his job by ferry. In the summer of 1916 he launched an editorial campaign for a Golden Gate Bridge. He lauded his extravagant praise of ancient and modern wonders and concluded that "the bridge across the Golden Gate would dwarf and overshadow them all."

Wilkins, who had studied engineering at UC Berkeley, painted a vivid picture of his suspension bridge that would run between Lime Point on the Marin shore to Fort Point in San Francisco. His design, a 3,000 foot center span with side spans of 1,000 feet each would cost about $10 million, he figured.

Wilkins' articles captured the attention of M. M. O'Shaughnessy, the San Francisco engineer who had energized the city's Hetch-Hetchy project which would dam and transport water from a valley adjacent to Yosemite, 156 miles to the rapidly growing city. During World War I O'Shaughnessy sought the counsel of engineers who deemed the project too costly with not much promise of reward.

However, one of the engineers contacted by the San Francisco Engineer was Joseph Strauss of Chicago, a graduate of the University of Cincinnati, whose graduation thesis (in 1892) was for the design for a structure to connect the continents of North America and Asia with a 50-mile bridge across the Bering Straits. He was also chosen as the class poet. Early-on Strauss implied he was an engineer, but historians have uncovered the fact that the University of Cincinnati didn't have an engineering college until 1901, nor graduate its first engineer until 1905.

But Strauss had vision and he had flair. He was also a man who made challenges welcome, who invented dreams to conquer them, and whose gifts would enable him to win the political, financial, and legal battles that lay ahead, battles perhaps more difficult than those posed by the winds and currents of the Golden Gate itself.

Strauss had gained fame — and money — with his re-design of bascule bridges. These were essentially similar to the draw bridges over castle moats, using either a single- or double-leaf span that would see-saw when its giant counterweight(s) were raised or lowered.

It's most likely that O'Shaughnessy and Strauss met during preparations for the 1915 Panama-Pacific Exposition, where Strauss designed the "Aeroscope," a popular ride. History has not marked the exact meeting of the pair, but does record that On June 25, 1915, the city of San Francisco awarded a contract for removal of the old swing drawbridge at Fourth and Channel Streets. Further, O'Shaughnessy's Department of Public Works designed the foundation for the new bridge over Islais Creek while the bridge was prepared by the J. B. Strauss Company of Chicago.

The Fourth Street Bridge, located in an industrial section of San Francisco is still in use today.

Another event occurred in 1915 that captured San Francisco's attention — the opening of the Panama Canal. The city seemed poised to become an important world center of travel and trade.

## Cars & Ferries

By 1920 there were more than a half-million automobiles in California and thousands more were being registered every month. Apparently one of the most popular destinations in the Bay Area was Marin and other counties to the north. Today it seems impossible, but in 1928, over two million cars were ferried across the Golden Gate.

## The Bank of America

The Bank of America maintains one of the best private archives of California History in the state. These archives contain material directly relating to the early financing of the Golden Gate Bridge. The following is their official version of some of the events leading up to the commencement of construction in 1933. As might be expected, Bank of America's story centers on its founder, A.P. Giannini and his personal contributions, but the story also captures something of the spirit of the times, an era when rugged individualists managed companies with their eye on more than just the bottom-line.

The following is from the archives of the Bank of America, and is used with their permission.

# THE BANK AND THE BUILDING OF THE GOLDEN GATE BRIDGE

*T*he beginnings of the Golden Gate Bridge project go back to 1918 when the San Francisco Board of Supervisors endorsed the idea of a survey of the proposed site to determine if a bridge across to Marin County might be feasible. The survey, completed in 1920, stated for the first time that a bridge was not only possible but practical. In January 1923, farsighted citizens on both sides of the Golden Gate came together to form the "Bridging the Golden Gate Association." By May their lobbying efforts encouraged the State Legislature to establish the Golden Gate Bridge and Highway District.

Because the proposed site of the structure (on both sides) lay on military property, and because gun emplacements and barracks there would have to be moved to construct the bridge, formal permission from the War Department had to be secured prior to any work commencing. The military needed to be convinced that the proposed structure would not be a menace to navigation and that it would be adequately financed. After formal hearings, permission was provisionally granted in December 1924. Throughout all this, the consulting engineer, Joseph Strauss, maintained that the estimated cost would be about $27 million. Few people believed him.

With the War Department's permit, now began a series of public hearings mandated by the Bridge Act to give taxpayers in the designated district area a chance to withdraw their individual properties from the tax rolls to fund the project. There followed six years of hearings, taxpayer suits, and county supervisors approving and then rescinding inclusion in the district. North coast timber, dairy, environmental interests, and the railroads (who controlled the trans-bay ferry business) strongly opposed the formation of the district. They produced engineering and geological experts who ridiculed the cost estimates and flatly asserted that the project was impractical, hairbrained, and that it could never be built. By 1929 over 2300 suits had been filed to stop it. Yet the district prevailed in all, and in December 1928 it was formally incorporated.

*From the archives of Bank of America, NT&SA.*

To begin construction, a bond issue would have to be approved by the voters of the new district. A well-organized and financed campaign soon sprang up against the bonds. The Citizens' Committee Against the Golden Gate Bridge Bonds began a series of radio broadcasts, paid advertisements, mailers, and door-to-door canvassing to arouse the voters. They stated the bridge was impossible to build. If built it would: mar the beauty of the site; destroy Sausalito's "splendid isolation" by being overrun with hoards of weekend picnickers; be an earthquake hazard; and if destroyed by an enemy fleet in time of war, it would bottle up our fleet in the harbor.

Local and prominent engineers urged defeat of the bond issue. A prominent banker flatly stated that to build the bridge during the Depression was an economic crime. It was also claimed that Marin's sparse population would never generate enough income from tolls to support the bridge. And the loudest argument was that the stated cost was grossly underestimated.

The bond election was held in November 1930 and passed by more than the required two-thirds majority. It authorized the sale of $35 million (face value) of bonds at a maximum rate of 5%. Revenue was also to be raised by property taxes. Construction bids were solicited and received in the amount of $24 million, many of them contingent on the sale of the bonds. Bankamerica Company, a bank affiliate through the holding company Transamerica Corporation, organized a syndicate and successfully bid for the bonds.

At this point, new litigation instigated by Southern Pacific-Golden Gate Ferries arose in the State Supreme Court that challenged the district's power to levy taxes. By the time the district won the case in November 1931, the bond syndicate's deadline had passed and the district had to borrow over $30,000 to pay salaries, fearing a new attempt to levy more taxes on the already unhappy taxpayers of the district.

Three days later, a challenge by two north coast firms, closely identified with Southern Pacific (Railroad), began new litigation in the U.S. District Court that the Golden Gate Bridge and Highway District be permanently enjoined from selling any bonds or moving in any way to build the bridge. This so aroused public indignation that a boycott of Southern Pacific began. The injunction was denied in July 1932, and Southern Pacific finally gave up its opposition to the bridge in August.

*From the archives of Bank of America, NT&SA.*

Bankamerica Company organized a new syndicate and again its bid of 92.3 for $6 million of bonds was accepted. However, at this point, the district's New York counsel was of the opinion that the effective interest rate of 5-1/4% exceeded the permissible rate. This was an issue that had to again be decided by the courts. Meanwhile, the Depression had deepened and the already poor bond market became worse. There seemed to be no solution to the problem, and it looked as if the bridge project would be indefinitely postponed.

*"We'll buy the bonds."*

In the late summer of 1932, Chief Engineer Strauss, together with a delegation of the district directors, decided as a last resort to call upon A.P. Giannini (left) and the Bank of America for assistance. Strauss had heard that A.P. was generous in support of anything relating to the betterment of California. They presented the district's past problems and present predicament and appealed for help. A.P. listened closely to their plea. He than called an assistant in and they briefly investigated the matter. He is then purported to have exclaimed, "California needs that bridge! You build the bridge, we'll buy the bonds."

Through the efforts of bank president Will F. Morrish, Bank of America reorganized the syndicate composed of itself, Bankamerica Company, Blyth & Co., Dean Witter and Co., and Weeden & Co. to purchase the first issue of $6 million of bonds. The bank agreed to immediately purchase $3 million of 4-3/4% bonds at 96.23 with an effective rate of 5%. This was done to protect the syndicate's previous bid of 92.3 for $6 million, pending its legality as decided by the courts. In addition, pending the determination of this question, the bank immediately purchased the first $200,000 (face value) of the invalidated bonds to meet the immediate expenses of the district. On September 2, 1932 Will F. Morrish delivered a check for $184,600 representing the advance sale of these first $200,000 of bonds. With this money, construction was finally able to begin in January 1933. The syndicate continued to purchase all of the district's subsequent bond issues as well.

*From the archives of Bank of America, NT&SA.*

The bank's action was taken in spite of the fact that 1932 was perhaps its most difficult year. In February, A.P. had emerged victorious from a vicious proxy battle to resume control of Transamerica and the bank. Deposits had fallen off drastically the previous year and the bank had to borrow and reserve large amounts from the Reconstruction Finance Corporation. To combat the Depression, the bank mounted an expensive "Back to Good Times" campaign during the year urging confidence and proclaiming that prosperity for California and the nation would be forthcoming. After the campaign began to make some headway in building up deposits, the end of 1932 saw the return of a steady deposit drain due to the beginnings of the national banking crisis in the east and midwest.

In the midst of these difficulties, the bank was recognized as doing something to bring about "good times" rather than just talking about it. The construction project put thousands of men to work and was represented as an unmistakable step away from the Depression and towards recovery and progress.

Chief Engineer Strauss later testified that without the bank's support for the bridge at that critical juncture, it would never have been built.

*From the archives of Bank of America, NT&SA.*

**At the Beginning: San Francisco.**

*What kind of city was San Francisco when work began on the Golden Gate Bridge? It was 27 years after the deadly earthquake and fire of 1906, and four years after collapse of Wall Street. In the midst of The Great Depression, San Francisco stood in splendid isolation, with no bridges to join her to the east or north. Despite the heroic rebuilding of the city following the disaster of '06, thousands were out of work, and the economy was stagnant. Lesser men might have decided to wait for better times before undertaking a massive "impossible" construction project. Joseph Strauss was not one of those men.*

# 1933

# WHAT WAS HAPPENING IN THE WORLD DURING CONSTRUCTION OF THE GOLDEN GATE BRIDGE?

**Historical Perspective: A capsule review of world events that occurred as the world famous span was being built.**

## Nineteen Thirty-Three

**Current Events.** Banks close in the U.S. Inauguration of F.D. Roosevelt, 32nd President of the U.S. He declares, "The only thing we have to fear, is fear itself," announces his 'New Deal.' Wiley Post flies around the world in 7 days, 18 hours, 49 minutes. Germany withdraws from League of Nations. In U.S., 21st Amendment repeals prohibition. Bauhaus design school closes in Germany. **Music.** Bartok, Piano Concerto No.2. George Balancine founds School of American Ballet. Richard Strauss, Arabella (opera). **Literature.** Erskine Caldwell, God's Little Acre. Andre Malraus, La Condition Humaine. Gertrude Stein, The Autobiography of Alice B. Toklas. **Births/Deaths.** Calvin Coolidge dies. Philip Roth, Richard Rogers, Akihito born. **Film.** King Kong, director Merian C. Cooper; Duck Soup, starring the Marx Brothers; The Invisible Man, starring Claude Rains; 14 Juli, director Rene Clair; Viktor and Viktoria, director Reinhold Schunzel. **Sports.** First all-star baseball game. AL defeated NL, 4-2, at Cominskey Park, Chicago. First NFL championship playoff. Gene Sarazen won PGA. **Everyday Life.** Ritz crackers introduced is U.S.

In Marin County, California on January 5, 1933, the first spade of dirt was turned to mark the commencement of construction of "the bridge that couldn't be built."

The first project had actually been to build an access road from the Waldo Grade to Lime Point where two large steam shovels would begin excavation needed for the Marin anchorage.

# WHERE TO BUILD THE BRIDGE EXACTLY? STRAUSS' COMMON-SENSE PLANNING

*After approval and Prior to the commencement of construction is that vital time when planning and details must coalesce. The Dream must merge onto paper — plans must include everything from the esoteric geometry of stress, to the chemistry of materials, the specifications, design and manufacturer of components, the acquisition and allocation of manpower . . . thousands and thousands of details, large and small. In other words, before a spade of dirt is turned, someone knows the answer to the question, "How many rivets in the Golden Gate Bridge?"*

*Before all the planning can be done, you have to know exactly where the Bridge will be located. These notes and others that follow in quotation marks, are from those who know best — Joseph Strauss and Clifford Paine and are contained in report of the chief engineer, issued upon completion of the Bridge in 1937, reprinted in 1987, and used with permission of the copyright owner, The Golden Gate Bridge, Highway and Transportation District.*

*So how to decide? His answer is straight forward.*

". . . because physical conditions at the entrance to San Francisco Bay are such that a bridge joining the two peninsulas should unquestionably be located at the Golden Gate, extending from Fort Point on the South to Lime Point on the North. Obviously any other location would involve a much longer structure and would introduce additional construction difficulties with no compensating reduction in those present at the site favored.

"From the south shoreline, the floor of the strait slopes gently toward the channel for some 1300 feet but beyond that the water depth increases rapidly. The foundation for the South Pier, desired to be not less than 20 feet into the rock, could be prepared under air pressure at a depth of 100 feet below the surface of the water. Meeting the above prerequisites placed the pier 1100 feet out from the shore where the maximum water depth at the channel face of the pier is about 80 feet.

"On the north shore, the steep slope of the rock bottom prohibited the placing of piers any appreciable distance off-shore. The length of the main span was thus fixed at 4200 feet since the water depth would make the construction cost of piers between these points prohibitive. The natural location of the South Shore-end Pylon at the Fort Point shoreline fixed the length of the side spans at 1125 feet. The narrow point of land which served so well to reduce the length of the structure left little leeway as to the location of the bridge in the east-west direction, but fortunately it was possible to keep the south cable-anchorage on land along the west shore of the Point where a setting in serpentine rock could be prepared for it. On the Marin side the 1125 foot sidespan had to skirt the high cliff and terminate in a Pylon just in front of the north cable-anchorage."

*Joseph Baermann Strauss.*

The wisdom of these choices will soon be evident.

"The anchorage at either end of the bridge comprises two separate masses of concrete into which each of the cables is anchored. Each mass was constructed in three units: the base block or foundation, the anchor block containing the anchor chains and girders; and the weight block which rests on top of the anchor block. The Marin Anchorage is located against the side of a sandstone and shale hill. The bottoms of the East and West base blocks are at elevations 78 and 84.5 feet respectively. On the Marin Anchorage, excavation was done by means of a power shovel and blasting powder."

**January 5, 1933.** Almost 14 years after Joseph Strauss selected the site, and 26 months after voters approved financing, a pair of giant steamshovels began excavation on a Marin County cliff at Lime Point to create the pit that would house the massive concrete blocks of the north anchorage. Construction of the Golden Gate Bridge was underway. Construction would continue without interruption until the span was opened May 27, 1937.

This modest beginning of a great engineering project was all Strauss would see for a while. After many years and more battles than he cared to remember, his physical and spiritual health drained, the Chief Engineer withdrew to the East to rest for six months, leaving Clifford E. Paine to manage construction. □

**San Francisco Anchorage.**

*Construction underway with Fort Point in the background. Anchorages are essentially huge concrete monoliths that would lock together with the earth and provide unmovable tie-downs for the two ends of the main bridge cables on each shore. This hole took longer than the Marin anchorage because it required blasting through serpentine rock. Also, the presence of Fort Point required some care and planning, which included removing part of the sea wall and replacing it after construction.*

## Aerial View, South Anchorage, Fort Point.

*The size of the excavation is difficult to judge, but contrasted to the trucks and steamshovels (center left), some idea of scale may be noted. This is also the site of the south pylons and the Fort Point Arch, a late add-on that allowed Strauss to save the historic landmark military installation whose purpose was to guard San Francisco Bay against invaders.*

BRIDGE UNDER CONSTRUCTION 1933

## Readying for Concrete.

*Forms were built to form three interlocking concrete forms. The bottom blocks were each 170 feet long, 60 feet wide and 97 feet high. Next would come the anchor blocks that would contain the tiedowns or eyebar chains, to which will be attached the 61 strands of the main cable. On top of this will come weight blocks to further immobilize the concrete mass.*

## The Marin Tower Pier.

*Strauss: "Design of the Marin Pier, located on the shoreline as it was, involved no noteworthy engineering problem. A U-shaped cofferdam built out from the shore about the pier site would permit excavation of the rock and preparation of the pier foundation at elevation -20 and subsequent construction of the pier, to be carried on as a land operation. The pier base was made 90 feet wide and 160 feet long, the length being determined as the minimum which would accommodate the base dimensions of the tower and the width as the minimum required for stability."*

*While the pier on which the north tower will be constructed seems massive, there is only a hint of the massive tower that will rise from it. In ensuing photos, watch how the buildings beside the pier seem to grow smaller as construction continues.*

*Building the Marin Pier was relatively easy. With 90 men working, the pier was completed in 101 days at a cost of $436,000. On the afternoon of June 29, 1933, it was turned over to the District, making it the first completed unit of the Bridge.*

## The North Tower Rises.

*The 746-foot towers are easily the most dramatic elements of the Bridge. Each tower contains 22,000 tons of steel. Despite their dramatic size, erection of the towers was one of the easiest jobs undertaken during construction. The steel for the towers was rolled and fabricated in two eastern cities, Pottstown and Steelton, Pennsylvania, where entire sections were assembled and shipped to Philadelphia, then by ship through the Panama Canal to a warehouse in Alameda.*

*This photo reveals how the towers were constructed by assembling "cells" of varying sizes. Note the*

*pre-fabricated sections being unloaded from the barge moored dockside, and the stacked units awaiting placement on the rapidly-forming tower. An unwieldy-looking "creeper derrick" moved upward as sections were completed. The flexible towers are made of 7/8-inch steel plates.*

*The lower transverse X-bracing, half-completed here, is used below the roadway. Above the roadway, horizontal portal bracing is used. The lack of X-bracing above the roadway is one of the distinctive design elements of the Bridge, giving the towers their clean, slender appearance.*

### Reaching Skyward.

*The Marin Tower rises above the roadway level. The creeper derrick has ascended several levels and the first horizontal bracing is in place. The roadway level is just above the uppermost "X." Bottom portions of the tower are being painted as work progresses. The access road leading to Sausalito is clearly visible.*

*Pre-fabricated sections are joined by white-hot rivets sent through pneumatic tubes. The first worker problems are caused by factory-applied red lead covering the steel. Hot rivets struck the paint, causing fumes. At first riveters wore gas masks, but despite this safety measure, 17 were struck by lead poisoning. This situation was cured when workers began removing paint from around the bored holes. Later, the sections were no longer painted with red lead when they were fabricated.*

# 1934

# WHAT WAS HAPPENING IN THE WORLD DURING CONSTRUCTION OF THE GOLDEN GATE BRIDGE?

Historical Perspective: A capsule review of world events that occurred as the world famous span was being built.

## Nineteen Thirty-Four

**Current Events.** Unemployment drops by more than 4-million workers. The first general strike in U.S. History takes place in San Francisco in support of 12,000 striking longshoremen. Only 58 banks failed this year, compared to an average of 901 per year since 1921. Securities and Exchange Commission established. Public Enemy Number One, John Dillinger, shot by FBI agents in Chicago. **Science.** Nylon invented by DuPont. 35mm Kodachrome film goes on sale. Cortisone discovered. **Music.** Rachmaninov, Rhapsody on a Theme by Paganini. First Oscar for Best Song, The Continental, from The Gay Divorcee. **Films.** The Black Cat, starring Boris Karloff. Cleopatra, dir. C.B. de Mille. Lives of a Bengal Lancer, Gary Cooper. The Thin Man. **Books.** James Cain, The Postman Always Rings Twice. F. Scott Fitzgerald, Tender is the Night. William Saroyan, The Daring Young Man on the Flying Trapeze. **Radio.** The Federal Communications Commission (FCC) formed. **Sports.** The U.S. Masters held for the first time in Augusta, Georgia. **Everyday Life.** Bonnie Parker and Clyde Barrow killed by police. U.S. Bureau of Prisons acquires Alcatraz Island in San Francisco Bay as site for a new federal prison.

**The Golden Gate Bridge construction** was speeding along. The north tower was beginning to rise from its base. For the first time it was evident *something* was being built, but as of now it showed little sign of becoming a graceful span across the raging gate.

### A New Shadow in Marin County.

*Amidst the headlands, the afternoon sun brings a new shape to the shoreline. Above the access road a section of the hill has been cut out to allow passage of the main cables to the anchorage, the cement blocks (right center). At far right is the concrete batching plant constructed to facilitate building the Marin Pier. The undeveloped area above the cutout is now an accessible view area.*

### San Francisco Anchorage and Pylon.

*This view shows the early 1934 development of the South Anchorage and the beginning of the twin concrete supports of one of the South Pylons. Despite the massive amount of work that included blasting and digging into the serpentine rock of the area, care was taken to protect every brick of Fort Point.*

*The South Tower would require more than twice the time and money of the Marin Tower. The overwater trestle leads from the shore to the location of the South Pier, 1100 feet from shore. Supported by piles driven into solid rock, the trestle proved to be the "bad luck" phase of early construction.*

### A View for a Bridge.
*North Tower construction speeds along. The creeper derrick sits atop the slender columns of the tower, lifting sections into place from the barely visible dock at its base. Painting has progressed about half-way up the tower. At bottom-center the work trestle stretches toward the location for the South Pier. Amidst the hectic pace of construction, Fort Point sits placidly, a monument to the past as a monument to the future is being formed around it.*

### Rising to Match the Mountains.

*The North Tower rises directly from the sea, seeming to rival the coastal mountains. The lighthouse to the right of the tower base now seems tiny. Access roads wind north toward the sleepy fishing village of Sausalito. There is no hint yet of the Waldo Grade highway, whose construction is the responsibility of the California Department of Highways. The out-of-focus "bar" on the right side of the photo is actually a wing strut of the open-cockpit aircraft used by the photographer taking this shot in February 1934.*

### A City of Bridges. It is June 1934.

This remarkable photo shows the Golden Gate Bridge (the North Tower, the San Francisco trestle), and the also-under-construction, Bay Bridge as it progresses toward Yerba Buena Island. The dark rectangle is Golden Gate Park with its clearly visible "Panhandle" and inside its boundaries, The Polo Field.

That the verdant park was literally created from sand dunes stretching toward the ocean is evidenced by the still undeveloped areas, lower right. The far hills of Alameda and Contra Costa counties reveal little development beyond the city of Berkeley. However, construction of the two bridges will vastly influence the future growth of the Bay Area to the East and North.

Alcatraz is the tiny, ship-like island, center. Missing from the photo: Treasure Island, a man-made landfill connected to Yerba Buena Island and built for the 1939 World's Fair and Exposition, later to become an airport for China Clipper flights and, until 1997, a U.S. naval base.

**From the Outer Harbor: Sitting Pretty.**

*The North Tower in June rises from beside the mile wide Golden Gate, only entrance to more than 400 miles of bay coastline. The channel, which appears benign in this photo, is victim of twice-daily tidal sweeps that can thrust 2,300,000 cubic feet of water per second through the channel at rates between 4.5 and 7.5 knots. The narrow opening to the sea is also one of nature's wind tunnels forcing gusts of up to 60 miles per hour directly across what will soon be the Golden Gate Bridge. These natural phenomena will affect the Bridge during construction — and decades later.*

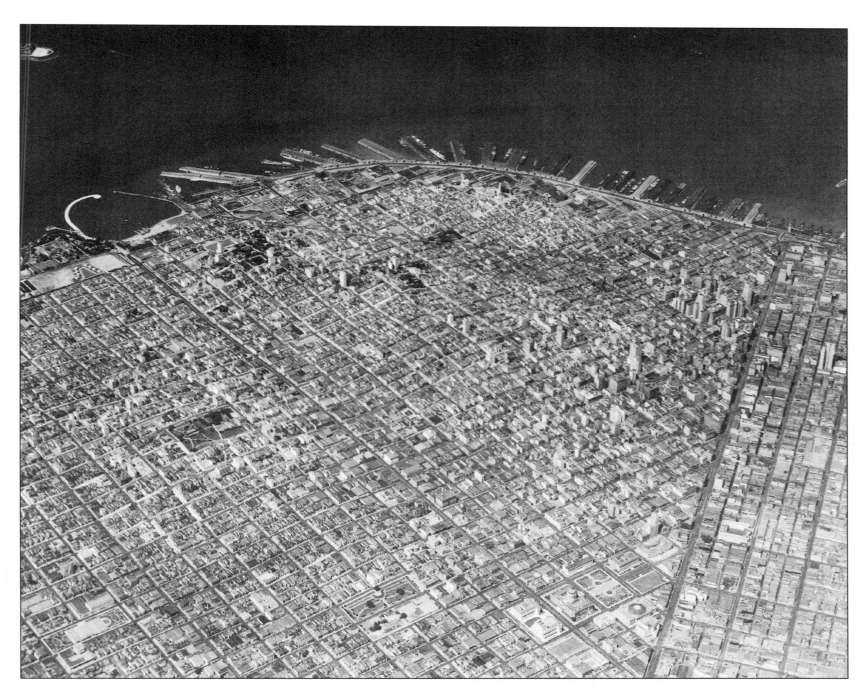

**The Cool, Gray City of Love.**
*San Francisco, except for a small cluster of office buildings in the Financial District (center, right) and a dozen or so tall apartment buildings, the city seemed to mold itself over its famed hills. At the foot of Market Street (diagonal, right) sits the Ferry Building. Van Ness Avenue crosses directly from the curved hook of the Aquatic Park pier to intersect with Market.*

**Peaceful, Graceful, Fast and Fragile.**
*Careful planning and scheduling has allowed the North Tower to rise quickly, providing an opportunity for bridge-watchers to begin to visualize how the Gate might finally be spanned. Few realized, however, that a suspension bridge is a highly interdependent structure, requiring careful engineering and construction to ensure stresses are harnessed and distributed uniformly to ensure the whole is stronger than the sum of its parts. The giant towers, for example, while masterpieces of design and engineering, are at their most vulnerable standing alone, without the benefit of cables and roadway to translate and modify stress forces. One observer of the time likened it to building one side of a four-sided building 746 feet high.*

### The Ring's the Thing.

Strauss: "The site of the San Francisco Pier is 1100 feet off-shore at the narrowest point of the Golden Gate where the tide sweeps with a velocity of 6.5 knots and where there is no protection from the full force of heavy seas and ocean storms."

Building the San Francisco Pier was the first time in the history of bridge construction that a pier was to be built in the open sea. Strauss' plan called for construction of a giant concrete fender to hold off the heavy surf and tides. After construction of the pier, the concrete ring would be retained to protect the pier from heavy seas and ocean-going traffic. Such a plan had never been attempted.

However, within days after this photo was taken the trestle was rammed in heavy fog by the outbound Sidney M. Hauptman, a vessel of The McCormick Line. The ship survived the accident, but tore away a 120-

foot section of the trestle. The damage estimate was $25,000, but the more crucial issue was that it would delay construction of the San Francisco pier by at least 30 days.

Tragedy was to strike the trestle twice more. On Halloween, October 31, 1933, another storm destroyed part of the trestle and portions of the sea wall fender, delaying pier construction another month. In mid-December, a southwest gale slammed in from the Pacific and took out 800 feet of the trestle and floated the twisted wreckage into shipping lanes. Predicted time lost: at least two more months.

However, storms on the surface weren't the only problem Strauss was facing. Below, in the murky, churning water where the blasting had taken place a new, even larger problem loomed.

## Strauss Meets with Board of Directors.

As the storm buffeted the trestle at the construction site, Chief Engineer Joseph Strauss was meeting with the District Board of Directors, and his report was far more serious than the destruction being raged on the trestle. As the storm bellowed through San Francisco, Strauss spoke directly to the issue at hand, revealing to a surprised Board that plans for the construction of the San Francisco Pier and fender would have to be completely redone. Moreover, he stated, the changes would cost $350,000 and could delay the completion of the bridge for 11 months.

Why?

Original plans had called for a concrete fender or wall the size of an oval football field, 155 feet wide and 300 feet long, with walls 27 1/2 feet thick. Inside the fender, which rested on an underwater base blasted from the serpentine rock at a depth of 65 feet, the plan was to excavate 35 additional feet inside the fender so that the pier would rest on bedrock at 100 feet.

However, blasting for the fender had shattered the rock base to such an extent it would be necessary for the fender as well as the pier to be taken down to 100 feet. The extra blasting and excavation would cost $10,000 per foot.

For a time, revelation of this situation created an outcry from several men who still opposed the idea of the bridge. One particular foe, an emeritus professor of geology from Stanford openly predicted that the foundation rock where the San Francisco Pier was to be constructed, "unstable to a degree likely to endanger the structure." In response, the Board of Directors of the District quickly hired an emeritus professor of geology from UC-Berkeley, who stated his Stanford counterpart's theory was "pure buncombe," adding, "No reputable geologist would consider it seriously." The issue ultimately faded away in a cloud of charges and counter-charges moving from geology into personal argument.

Work continued on the San Francisco Pier, the most troublesome element of the of the entire project. The president of the construction company that built the San Francisco Pier called it, "... one of the most difficult and hazardous pieces of construction work ever attempted."

The pier was poured by crews working day and night inside the fender walls.

## Yes, It is Very Large.

*The fender, looking inside. The men are standing on the pier itself. Below, water has been let into the area between the fender and the pier to equalize pressure. The cylindrical, capped-off pipes are eight "inspection wells" to allow access to the bedrock.*

*Construction of the San Francisco Pier and fender continued to a height 40 feet above the water. This photo from August 1934 shows a CBS radio reporter and his engineer viewing the nearly-completed pier prior to a broadcast.*

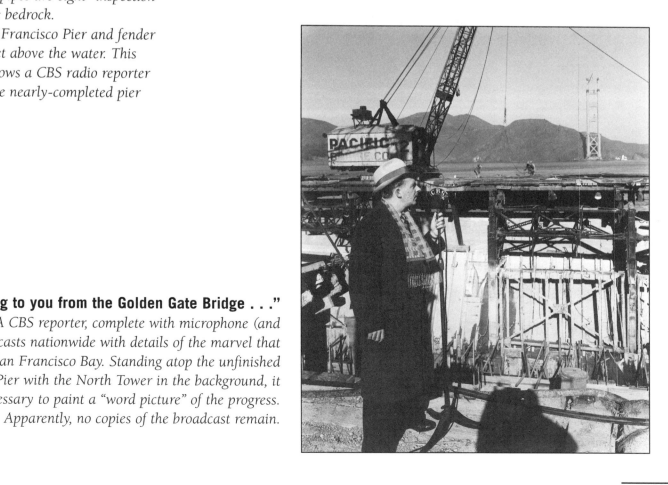

## "I'm speaking to you from the Golden Gate Bridge . . ."

*Well, almost. A CBS reporter, complete with microphone (and stand) broadcasts nationwide with details of the marvel that was rising in San Francisco Bay. Standing atop the unfinished San Francisco Pier with the North Tower in the background, it was necessary to paint a "word picture" of the progress. Apparently, no copies of the broadcast remain.*

**The South Tower: After the Concrete.**

*Once the pier was complete its surface had to be prepared to accept base plates to begin the tower construction. Strauss describes the process.*

"Preparation of the pier top to receive the tower base slabs was a part of the work included in the tower contract. The specifications required the top surface of the concrete under the tower base to be ground to a level plane. This was done with a carborundum grinding wheel mounted on a trolley. The trolley supporting bridge itself was arranged to move on rails accurately leveled. With this equipment the bearing area under each tower shaft, amounting to about 1200 square feet, was finished to the required level plane of less than one-thirty-second of an inch. As soon as the tops of the pier under the bearing plates had been surfaced, the base plates were set in a thick coating of freshly applied red lead paste. By varying the thickness of this red lead cushion, the plates were set to an accuracy of .003 inch. After the plates had been set, the steel dowels, 4 feet 6 inches long and 6 1/2 inches in diameter, were set in holes provided in the top of the pier and grouted into position."

# 1935

# WHAT WAS HAPPENING IN THE WORLD DURING CONSTRUCTION OF THE GOLDEN GATE BRIDGE?

Historical Perspective: A capsule review of world events that occurred as the world famous span was being built.

## Nineteen Thirty-Five

**Current Events.** The Works Progress Administration (WPA) organized. The Rural Electrification Administration (REA) formed to build power lines outside cities. The Social Security Act was signed by President Roosevelt. Persia changes its name to Iran. French trade unions recognize right of women to hold jobs. **Science.** British driver Malcom Campbell sets land speed record of 445.4 m.p.h. at Daytona Beach, Florida. SS Normandie crosses Atlantic in 107 hours, 33 minutes. American Charles Richter introduces Richter scale for measuring earthquakes. **Music.** George Gershwin, Porgy and Bess. Richard Strauss, Die Schweigsame Frau (opera). Popular song, Happy Birthday written. First Gibson electro-acoustic guitar sold. **Books.** John Steinbeck, Tortilla Flat. Andre Malraux, Le temps du mepris. **Films.** Fox merges with Twentieth Century Co. First of 66 Hopalong Cassidy films made. Anna Karenina, Greta Garbo. A Night at the Opera, Marx Brothers. Top Hat, Fred Astaire and Ginger Rogers. **Sports.** First night baseball game, Reds vs Phillies, Crosley Field, Cincinnati. **Everyday Life.** Alcoholics Anonymous (AA) organized in New York City. Radio program, Your Lucky Strike Hit Parade, begins. New board game, Monopoly, goes on sale.

**The Golden Gate Bridge** was humming along. With the problems and setbacks of the San Francisco Pier behind them, construction crews completed the concrete work and made ready to commence assembly of the South Tower. It would be a busy and productive year.

BRIDGE UNDER
CONSTRUCTION
1935

### San Francisco Anchorage.

*Forms, built by scores of carpenters await concrete as the second pylon rises to the height of Fort Point's walls. Between the pylons, designed to support the roadway approach to the bridge proper, a huge arch of steel girders will be constructed.*

*Early design of the Golden Gate Bridge called for the demolition of the historic fort. After the decision was made to retain it, Charles Ellis who worked with Strauss (and is today given much of the credit for the final design), came forward with a graceful solution for what could have been an aesthetic flaw, providing an artistic frame for the weathered brick of Fort Point.*

### Watch it Grow!

*The San Francisco Pier was finished on January 8, 1935. Work immediately began on the erection of the San Francisco Tower and by February (this photo) the tower legs were visible, along with the creeper derrick. Sections were unloaded onto a working platform on the channel side of the fender. The District worked out a bonus agreement with the contractor to speed construction of the tower in order to make up for some of the time lost with the trestle problems, and to move forward the opening date of the bridge. The job was completed in 170 days.*

**BRIDGE UNDER CONSTRUCTION 1935**

### Rat-tat-tat-tat-tat. Rat-tat-tat-tat-tat!

*In joining the pre-fabricated sections, all of the horizontal joints were riveted before the vertical seams were driven. There were approximately 600,000 field-driven rivets in each of the towers. The rivets were heated in coal-fired forges placed on scaffolds outside the tower. They were transported to crews inside the tower by 6 1/2 inch pneumatic tubes.*

BRIDGE UNDER CONSTRUCTION 1935

### Safety First.

*Strauss: "Special provisions were made for the safety of the men during the erection. Every man was required to wear a hard hat to minimize head injuries from falling objects. Because of the confined space within cells, extra precautions were taken against lead poisoning. This ailment was suspected in several workmen and as soon as its likelihood was discovered, all men were examined physically every two weeks and blood counts were taken. All riveters were required to wear respirators, and provisions made so that hands could be kept clean to prevent hand to mouth infection. It is believed that the erection of the two towers, involving 44,000 tons of steel without a fatal accident, established a noteworthy record."*

BRIDGE UNDER CONSTRUCTION 1935

53

### On a Clear Day You can See . . .

*March 1935. Thousands of San Franciscans could follow the building of the Golden Gate Bridge from their office or apartment windows. The distance between the towers, 4200 feet, still exceeded that of any suspension bridge ever built. However, now even the harshest doubters were beginning to visualize the graceful link that would join San Francisco and Marin Counties.*

BRIDGE UNDER CONSTRUCTION 1935

### Steady as she goes . . .

*By now merchant seamen from around the world who passed through the Golden Gate were beginning to take stories about the colossus that seemed to grow daily back to their home ports, where their "tall tales" doubtless fell on doubting ears. The same workers who built the Marin Tower were used to build the San Francisco Tower. Not a happenstance, this was part of Strauss' planning.*

**Full Speed Ahead.**

*Within a few weeks after completing the X-bracing beneath where the roadway would run, workers were installing the next-to-last horizontal braces between the tower legs. The designers of the bridge chose to emphasize the majesty of the towers by slimming the towers from the base to the top. The hollow, pre-fabricated cells measured 42 inches square and 35 feet in length. The base of each leg would begin with 97 cells, and narrow in stages, with the top containing only 21 cells. Strauss, who had final approval over every design detail, lauded Clifford Paine and Irving Morrow, boasted, "This bridge is perhaps the first in which the importance of the new motif of stepped-off towers has been recognized and applied."*

**Bird's Eye View.**
*The South Tower grows thinner as it climbs higher. Another design feature that helped lend a "doorway effect" was the sizing of the horizontal portal bracing above the roadway. Each succeeding brace decreases in size. The Golden Gate Bridge was also the first bridge to eliminate transverse, or X-bracing between the tower legs, lending to the lightness and majesty of the huge towers.*

BARRETT & HILP

**Getting Cars to the Bridge.**

*The approach from Lombard Street through the Presidio
began at grade and elevated in order to reach the
approach Toll Plaza area at the south end of the Bridge.
While this modern approach was being built, there was
still a problem of getting autos to this point. Lombard
Street had not yet been widened and as yet there was no
approach joining Funston Avenue. Early expectations
called for a Bay Street Tunnel that would connect the
North Beach area with this Bridge approach.*

BRIDGE UNDER CONSTRUCTION 1935

**So Long, Sailor.**

*Ships of the Pacific Fleet embark from the Port of San Francisco, April 1935. The Golden Gate Bridge would be a familiar symbol to American soldiers and sailors who would pass under it enroute to far-off duty.*

*Even before it was completed, the Bridge would begin to weave itself into the emotional fabric of America by becoming the Western Gateway to the United States. Ten years later, military wives would crowd the sidewalks of the Bridge and wave to returning ships, hoping to catch a glance of a returning husband.*

*In more recent times, the Golden Gate Bridge has come to symbolize the outward quest for international commerce and trade. As such, it is often referred to as "The Gateway to the Pacific Rim," a profound and humble motto of high hope and great expectation.*

**The South Tower Nears Completion.**

At 746 feet, the towers are one-third the height of Mt. Tamalpais, "the sleeping princess." They are 232 feet higher (and many times more aesthetic) than the towers of the also-under-construction San Francisco-Oakland Bay Bridge. The tallest building in San Francisco at this time is the Russ Building on Montgomery; these towers are twice its height. San Franciscans were beginning to understand the scope and challenge of the undertaking — and beginning to develop the civic pride that would come with the construction of the world's longest suspension bridge on their northern doorstep.

## The Twin Towers: Ready to Become a Bridge.

*This photo shows the clear path the Bridge's cables will take. From the top of the South Tower the cables will descend past the "notches" on the pylon on the Bay side of Fort Point. They will proceed from there through the door-like space in each leg of the pylon this side of the fort, thence to the waiting eye-bolts of the San Francisco Anchorage. The pylons, of course, support the roadway that will be built above and anchor the archway that will be built over the fort.*

*Approximately one month after this photo was taken an unusual event occured — the Golden Gate was closed to shipping, no boats were allowed in or out of the strait. On August 2, 1935, a heavy fog shrouded the north tower. At nine o'clock that morning a barge moved into the fast-running tide, heading from Marin to San Francisco. As it crossed, a 1 9/16-inch wire rope slowly unspooled behind. It took the boat an hour in the rushing current to make the crossing.*

*At noon, the derricks atop each tower began to raise the wire rope from where it lay on the bottom of the channel, 350 feet below the surface. In just eight minutes the wire was raised to the top of the towers. Many called it a miracle that it did not snag on rocks or submerged shipwrecks. The wire was allowed to create a gentle arc between the giant towers, and the ends were draped to each shore and fastened.*

*The land between the north and south shores of the Golden Gate was joined.*

**Big Day at the Office.**

*From their workplace atop the north tower, the derrick and wire crew hoist cables, make ready for construction of the footwalks they would need to spin the cable. While the day appears clear and the weather mild, the summer fog and winds were daily visitors thoughout the months it would take to create the support cables.*

**Smile for the Camera.**

*A moment of quiet satisfaction in a job well done. Photographers covering the construction of the Golden Gate Bridge seldom asked the workers to pose for photos and this is a rare exception. Note the hard hats as part of the safety equipment. Workers were also required to apply sunscreen to guard against the harsh environment at the top of the 746-foot towers. This photo was taken September, 1935.*

**BRIDGE UNDER CONSTRUCTION 1935**

**"Stringing the Harp of Heaven . . ."** *is how one person described the delicate-appearing work of installing the ropes that would hold the catwalks the wire workers would need. Taken from the top of the north tower, it is hard to imagine a bridge so sturdy that fifty years later it would hold more than 300,000 celebrants. Approximately twenty months after this photo was taken, the bridge opened to motor traffic.*

**Ships That Pass in the Day . . .** *now pass under the first wires to bridge the shores of the Golden Gate. Just above the uppermost "X" bracing is where the roadway will pass through the tower. During this era, San Francisco was the busiest seaport on the Western Coast of the United States and the twin towers of the bridge quickly became the remembered view of many a sailor entering or leaving the port.*

**Take a Good Look.**

*From water-level, this view is about to change forever.
In a few months the towers would be joined by giant
support cables, a roadway would drape gracefully
between them and the entrance to San Francisco
would be framed by the Golden Gate Bridge.
While the immensity and scope of the project may
not seem as significant today as it did in the nineteen
thirties, the men, machinery and materials of that
day resulted in a man-made marvel of the ages.
Engineering and aesthetics have seldom combined to
create beauty, utility and durability in such harmony.
The evidence of that could be sensed even at
this early state of construction.*

**Pathways of Steel.**

As the wire ropes are strung and anchored, you can now see the path the main cables will follow as they descend from the top of the tower, past the "notches" of the first pylon and through the "holes" in the second pylon to the anchorage where they will be tied to huge eye-bars sunken in the concrete and topped with weight blocks.

**Man Contemplating His Future.**
*This September 1936 photo is
one of the most famous shots of the
unfinished Golden Gate Bridge. A
lone man, tools in hand, stands on
two wire ropes, each about one and
one-half inches in diameter, atop
the north tower. Trecherous cross-
winds often howled through the gate
and men were forced to work in
heavy fog that limited visibility
sometimes to mere inches. A slightly
different version of this photo is
used for the cover of the book,
"Spanning the Gate", written by
Stephen Cassady, published by
Squarebooks, Mill Valley,
California.*

### Stairway to the Stars?

*Not quite. The cables supported these footwalks. Constructed from steel channeling and heavy redwood planks, they would run from shore to tower, arc across the water to the opposite tower and then to shore again. Spaces were left between the redwood planks to allow wind through and thereby remain more stable. Pre-assembled in 10-foot lengths, they were pulled up the wires with men on board who clamped them to the support cables.*

BRIDGE UNDER CONSTRUCTION 1935

**Seasonal Fog, Seasoned Men.**

*Fog brought its distractions, but no interruption of work. The pre-assembled footbridge sections for the tower-to-tower portions were hoisted to the tops of the towers from barges, then slowly allowed to slide down the support wires until they were clamped in place. Steel-truss "crosswalks" were created about every five-hundred feet. In addition, a series of guy wires were strung from the footbridges to the towers to help stabilize them and reduce movement due to cross winds. Note the worker (top, center) and his casual toe-hold as he works overhead.*

**September 27, 1935.**

*A worker on the San Francisco side waves across the first bridge to span the Golden Gate to his counterpart somewhere on the Marin side. The footbridges are completed and will remain in place for eight months while 80,000 miles of wire is spun between the two anchorages.*

### The Shadow Knows . . .

*At center, right, an unfamiliar shadow on the surface of San Francisco Bay, created by the temporary footbridges which will soon give way to the main cables, suspender cables and roadway of the Golden Gate Bridge. Details of the crosswalks may be seen, center. This 4200-foot span will be the longest suspension bridge in the world for many years to come. By the end of 1935, most of the doubters had become fans of the great orange bridge which would open to the public in less than eighteen months.*

**Entering a New Phase.**

*The tower contractor completed his work in mid-year and turned the towers over to the cable contractor who, upon completion of the footbridges and assorted equipment assembly, would begin forming the main support cables of the bridge. This operation, complex and not clearly understood by non-engineers, would be one of the most fascinating operations of the bridge's construction.*

**Optical Illusion.**

*Looking down toward the base of the north tower from just above the third lateral brace. The tower appears wider at the top because of the camera's closeness to it.*

*Many people who stare at the base of this photo claim to become dizzy. Vertigo was certainly no problem to the workers who scrambled around these towers daily.*

**December 1935.**

*The guying wires which helped steady the footbridge in high winds were decorated with red and green navigational lights to aid in ship navigation. Unexpectedly, the uncompleted bridge towers became nighttime Christmas trees for a city already enchanted with the prospect of a grand, new "impossible" bridge.*

**Flying High January 1936.**

*A Pan American Airways "China Clipper" soars over
the work-in-progress that will open in about 17 months.
The era of creative engineering that created the flying
boat and the Golden Gate Bridge would mark America's
world influence for years to come. As passengers relaxed
for their long (and noisy) trip to Hawaii, a team of
supervisors from contractor John A. Roebling & Sons
were readying wire crews for California's first Big Spin.*

# 1936

# WHAT WAS HAPPENING IN THE WORLD DURING CONSTRUCTION OF THE GOLDEN GATE BRIDGE?

Historical Perspective: A capsule review of world events that occurred as the world famous span was being built.

BRIDGE UNDER CONSTRUCTION 1936

# Nineteen Thirty-Six

**Current Events.** There are still 8-million unemployed in U.S., but auto production increased 20% and farm prices were up. Hoover Dam opens on the Arizona-Nevada border. In Germany, first Volkswagen produced. FDR reelected president. **Books.** James Cain, Double Indemnity. William Faulkner, Absalom, Absalom! Margaret Mitchell, Gone with the Wind. British publisher founds Penguin Books, begins the paperback revolution. **Music.** Sergei Prokofiev, Peter and the Wolf. Rodgers and Hart musical, On Your Toes, opens. Arthur Johnston, Pennies From Heaven. Swing Era opens, Benny Goodman named King of Swing. **Films.** Flash Gordon, starring Buster Crabbe. Mr. Deeds Goes to Town, dir. Frank Capra. Modern Times, starring Charlie Chaplin. Things to Come, prod. Alexander Korda. **Sports.** Joe Louis loses Yankee Stadium non-title bout to Max Schmeling in 12 rounds. In Summer Olympics in Berlin, U.S. athlete Jesse Owens wins four gold medals. The Yankees beat the Giants in the 33rd World Series, four games to two. **Everyday Life.** Everyone was reading Dale Carnegie book, How to Win Friends and Influence People. Movie attendance soars; studios turnout 500 features a year.

**On the Golden Gate Bridge,** towers are completed and the major task of "spinning" or making the main cables moves into high gear. Engineer Strauss comes up with another new work-safety idea.

### Strauss on the Main Cables

*"The cables are the largest built to date. Each of the two cables is thirty-six and three-eighths inches in diameter (over wrapping at a distance from the cable bands) and 7650 feet in length between the centers of the eye-bar pins at opposite anchorages. Each cable is made up of 27,572 parallel wires distributed into 61 strands. The great lengths of the spans, combined with certain unique weather conditions encountered at the site, introduced new experiences in cable construction. Cable construction was marked by several innovations in design and erection practice."*

# Why Did They Have to Manufacture the Cables at the Bridge?

Think about it. The weight of the Bridge main cables, including suspender cables to hold the deck, along with other accessories is 24,500 tons. It was (and still is) impossible to fabricate these cables on land and raise them 746 feet into the air and place them in saddles atop each tower. Then each of the sixty-one strands of the four ends of the two cables would have to be connected to the eye-bars that were embedded in the massive anchorages at each end of the Bridge. The only answer was to manufacture them in mid-air, and that required the genius of another man.

## John A. Roebling

Roebling, a German-born engineer emigrated to the United States in 1831, settling into a placid life of farming in Western Pennsylvania. However, he kept his professional interest in engineering and began to make wire rope in his spare time finding a ready market among canal barge owners who preferred his sturdy product to the then-popular hemp.

Soon he gave up farming and devoted the balance of his life to the construction and engineering of wire rope. His fame spread rapidly and his wire was used in construction of the Brooklyn Bridge, completed in 1883. This project marked the first time wire rope was protected with a coating of galvanized zinc. The 15-1/2 inch cables of the Brooklyn Bridge were spun at the site using a special invention created by Roebling — a mobile spinning wheel-carriage. It was this invention that would unspool a loop of wire as it shuttled back and forth between San Francisco and Marin.

### Here's How It Worked.

*There is probably more confusion and misunderstanding about the spinning operation than any other part of the construction of The Golden Gate Bridge. As with all great ideas, the method was startlingly simple.*

*As soon as those footbridges of steel girders and redwood planks were completed, a set of hauling cables was installed above. These would carry the spinning wheels from shore, up and over the first tower, down and then up the graceful swoop between the towers, over the second tower and down to the shore for anchoring. A wire would be looped over the wheel and as it travelled, it pulled wire from a reel on shore.*

***If you still can't visualize the way it worked, try this:*** *Tie a long piece of string to the back of a chair or to a doorknob. Let the string dangle on the floor. The pile of string on the floor represents the spool the wire is on. Stand sideways to the chair or doorknob, grasp the string in your left hand and pull it out a couple of feet. Hold the string and cross your right hand over your left and put the pencil under the string between where you are holding and where it is tied. Now lift the pencil only upward, and to your right, keeping your other hand in place. Allow the string to pass through your fingers.*

*What you're now doing is pulling a loop that's feeding smoothly through your fingers — **just the way Roebling's simple invention pulled the wire to make the Golden Gate Bridge Cables.** The pencil, of course, is the carriage with the wheel over which the wire passes.*

*The carriages did not make the full crossing. Instead, two of them on the same cable would start from opposite sides. When they met in the center, the wires would be exchanged, so that each carriage returned to its starting place, completing its half-way-then-return journey with the wire from the opposite side. Simple, efficient and fast. How fast we'll see shortly.*

*The final element, of course, was joining or splicing the individual wires end-to-end. There are 117,000 such splices in the wires that make up the main cables of the Golden Gate Bridge.*

*Finally, each wire had to be adjusted to conform to the guide wires which had been precisely-engineered to "sag" across the Gate. The sag was exactly 475 feet and each of the spun wires had to conform to the guide wires in order to ensure precision with the mathematical computations of the bridge structure.*

### Complicated Geometry.

*The slim towers require great care when stresses are applied. The metal "saddles" over which the main cables run were each set up with a heavy base plate containing openings for thirty-four steel rollers. These rollers, eight inches in diameter weighed more than three-quarters of a ton each. As the main cables are formed, suspender cables and roadway added, the saddles would move on these rollers. Their design was such that when the bridge was completed, the saddles would be in the center of each tower. Following that, the cables would be permanently locked into place in each saddle.*

### What this means.

*Permanent fastening of the cables meant that in the case of changes in ambient temperature, wind, bridge load and other factors, the cables would not slide in their saddles. In the event of any of these factors, their resulting movement would be transferred to the towers themselves. The engineering dictated that the tops of the towers be able to lean twenty-two inches shoreward and eighteen inches toward each other, a total movement of forty inches. The towers are also engineered to lean 12-1/2 inches from side to side.*

**The Transfer at Mid-Span.**

*This photo probably proves the old maxim that not much gets done when someone is taking your picture. The standing-waiting appearance of the men is contrary to the facts. History shows, in fact, that the cable-spinning operation was a smooth, fast, untroubled element of the Bridge's construction.*

**Band That Strand.**

*Workers band individual wires together every five feet with heavy-duty tape to make a strand. Each strand contained an average of 452 wires. Sixty-one of these strands make up the main cable. Each strand is tied to an embedded eye-bolt to transfer the pulling force of the cables to the anchorages. The waters below appear choppy and wind was one of the biggest problems encountered during the spinning operation. At one time 40 MPH winds swept through the Gate, swinging the catwalks nearly eight feet from where the cable was being spun. The spinning operation continued throughout the storm.*

*The corregated tin structure (center, left) is a "warmup shack" where men could get out of the wind a few minutes to warm up. Regular times were allotted for men to use the shacks. By today's standards, they do not appear to be much help, but old-timers claim they were beneficial.*

BRIDGE UNDER CONSTRUCTION 1936

**South Tower to the Ground.**
Once the strands were completed they were loosely
formed together (center). The outside strands were
banded and then moved to the proper location. Note the
strands once again crossing the "notch" in one pylon and
passing through the "windows" of the second pylon,
then to the eye-bolts of the anchorage.

**One Side at a Time.**
*The work of adjusting tension and sag over the tower saddles meant no work could be done on that tower. Action moved to the opposite tower where spinning would proceed. When the strands were adjusted on the first tower, spinning action would resume there while saddle work was performed on the second. Thus it went, back and forth until the cables were completed. Shown here: The west cable with completed strands bolted into separators to keep them clear as other strands were being spun and adjusted.*

**Fine Tuning.**

*As the strands passed over the saddles of each tower, they were adjusted for tension and sag by means of the U-shaped lifting straps and seizing bands seen here. The strands have been placed and workers are cutting away the straps.*

BRIDGE UNDER
CONSTRUCTION
1936

**Meantime, Back at the Approach . . .**

*With the cables making progress, supports for the approach road marched across the Presidio of San Francisco landscape in early 1936. There is not much information regarding problems incurred with road construction, leading to the assumption this part of the immense project was completed with relative ease.*

## Coming Up: The Big Squeeze.

*When the last wire had been spun the cables resembled a hexagon nearly five feet in diameter. What remained now was to squeeze the cables together to make them very nearly round and give them the strength of a single unit.*

*Roebling's spinning machine underwent a major change after 122 strands had been spun. An additional wheel was added. After a short trial period, workers were stringing 271 tons of wire a day. With this improvement, 80,000 miles of wire was spun for the main cables and the spinning was completed on May 20, 1936. This date was nearly two months ahead of time and exactly one year and one week before Opening Day.*

**Compacting and Wrapping.**
Here's Strauss' explanation of the process. "Six compactors were used to squeeze the cable to a circular cross-section. Each consisted of a frame which surrounded the cable and supported a battery of twelve jacks radially directed so as to exert a force normal to the surface of the cable."

**Strauss Continues:**

*"The jack assembly was supported on a structural steel frame which was designed to travel on the cable on large wooden spools. The frame was stationary for three moves of the jacking assembly which rode on tracks on the frame. After the three moves, the jacking assembly gripped the cable and the supporting frame was moved ahead to a new position in readiness for three more moves of the jack assembly."*

**Strauss:**

*"All movements of the assembly were made by means of hand-operated cable winches. Of the jacks having a piston diameter of 5-7/8 inches and operating under a maximum pressure of 6000 pounds per square inch, exerted a pressure of 81 tons, or a total of 972 tons for the twelve jacks in each compactor. The shoes contacting the surface of the cable were 6 inches wide and completely surround the cable so that the total contact area was about 678 square inches.*

*"After the start of operations, it was found that little further compaction was accomplished after the fluid pressure in the jacks exceeded 5000 pounds per square inch, and for most of the work, this pressure was used."*

Note: The small nodule in the center is a splicing ferrule. There are 117,000 splices in the wires which make up the main cables.

**Keeping the Cable Compressed. Strauss writes:**

*"Incorporated in the compacting unit were two auxiliary hydraulic jacks which tightened the seizing bands. The cable was compressed at intervals of three feet, which therefore was, in general, the seizing band spacing. The seizings consisted of galvanized metal straps, two inches wide, which were held tightly by engaging the ends in a special socket and wedge device. After the adjustments under working conditions had been made, each compacting machine was able to compact 204 lineal feet of cable per day of eight hours."*

**Check It Once . . .**

*Following closely behind the compacting machines, a supervisor checks the squeezed diameter of a main cable. According to Chief Engineer Strauss, "The mean diameter after compacting was close to 36-1/16 inches (at the sizing bands). Under the permanent bands, after final tightening, the mean diameter was 35-7/8 inches."*

**. . . and Check It Again.**
*After being compacted, each of the main cables of the Golden Gate Bridge was tightly wrapped with galvanized wire to provide a protective coating. This was done with six wrapping machines, also developed by Roebling.*

In Case You Were Wondering.

BRIDGE UNDER CONSTRUCTION 1936

**The Binds that Tie.  Strauss again:**

*"The suspenders consist of four parts of 2-11/16 inch diameter wire rope. The longest suspender rope is approximately 980 feet long overall. They were unreeled from the tower tops, both ends being hauled down the footbridge by use of the high-line, and then lowered over a pair of temporary sheaves mounted on the cable band. A center mark on each rope was made to correspond with the top of the split in the cable band."*

**Happy But Not Connected.**
*The suits may have waved and shouted, but none of them was about to attempt crossing the girders to shake hands with their counterparts on the opposite side. After the waving and shouting, they went home. Strauss is seated center.*

**The Dream Becomes A Reality.**

*With virtually all of the steelwork completed, the Golden Gate has been bridged. Much remains to be done, paving, lighting, sidewalks, guard rails, the toll plaza . . . But here, in its bare-bones, the promise has been realized. According to Strauss: "Substantially all suspended steel was erected on December 14, 1936.*

BRIDGE UNDER CONSTRUCTION 1936

# 1937

# WHAT WAS HAPPENING IN THE WORLD DURING CONSTRUCTION OF THE GOLDEN GATE BRIDGE?

**Historical Perspective: A capsule review of world events that occurred as the world famous span was being built.**

## Nineteen Thirty-Seven

**Current Events.** Franklin Roosevelt begins second term. Hundreds of Americans volunteer in the Spanish Civil War, join Loyalist armies. Minimum wage law for women upheld by Supreme Court. The dirigible Hindenburg destroyed by fire as it attempted a mooring at Lakehurst, NJ. The disaster ended lighter-than-air transport. The first coast-to-coast radio broadcast featured Herbert Morrison describing the disaster. In France, Duke of Windsor, formerly King Edward VIII marries Mrs. Wallace Simpson. **Science.** British engineer builds first prototype of jet engine. In U.S., C. F. Carlson pioneers Xerography. **Music.** Shostakovich, Symphony No. 5. The NBC Symphony Orchestra, with conductor Arturo Toscannini as conductor is established. Bing Crosby has first hit record. Woody Guthrie stars in local radio show on KFVD

in Los Angeles. **Films.** Camille, starring (cough, cough) Greta Garbo. Snow White and the Seven Dwarfs. Prix Louis Delluc founded for best French film of year. Lost Horizon, dir. Frank Capra. La Grande Illusion, dir. Jean Renoir. **Books.** John P. Marquand, The Late George Apley. John Steinbeck, Of Mice and Men. **Everyday Life.** NBC begins regular experimental television broadcasts. U.S. spinach growers erect a statue to Popeye. Major floods in Midwest leave a million homeless.

On the Golden Gate Bridge, work is wrapped up, a celebration is planned, the excitement grows. The "impossible bridge" is about twenty weeks from opening. However, one of the major catastrophes of the construction was to take place only seven weeks into the new year.

**Roadway Pavement and Sidewalks.**

*At the beginning of 1937 crews readied the roadway for paving. Installation of the sidewalks would follow and finally removal of the footbridges which had been used during main cable spinning and suspension cable hanging. Painters were still working on the suspender cables as well. Reinforcing steel in two 20-foot strips for the first-phase of concrete work may be seen (center). Strauss had planned for the paving contractor to begin work before the steel work was completed in order to save time, but problems between the contractors delayed this.*

### A Pair of Johnnies On-the-Spot.

*The reinforcing mesh was spot-welded in two-man teams. Each weld was inspected and after being approved, concrete work could begin. All of this field welding was done by portable gas-driven welding machines hauled onto the unfinished roadway.*

BRIDGE UNDER CONSTRUCTION 1937

**Low Tech At a High Point.**

*Painting of the suspender cables is by time-tested brush-and-bucket technique. Painters, working in bosun's chairs for long hours, survived wind and sun, but were bothered most when their legs fell asleep.*

**Ribbons of Steel Joined by Ribbons of Concrete.**
*The two outside slabs of roadway would be poured first. These 20 by 50 sections would later be filled in when the center slabs were poured. Other finishing touches were underway. Note the footbridge is removed from beneath the west main cable yet remains on the east cable — for now.*

**Sidewalks and Guard Railings.**

*The sidewalk, guard rails being installed in 1937. In future years, consideration would be given to replacing the solid sidewalks with grates to cut down wind resistance. In 1993 these rails on the west side of the Bridge were removed and replaced with exact duplicates.*

**BRIDGE UNDER CONSTRUCTION 1937**

**The Waldo Grade Being Graded.**
*This arch is the single-bore tunnel through which the highway to and from the new bridge will pass. For the next few years, the District will press for an alternate route through Sausalito because the steep grade was a difficult encounter for the trucks of the day. The second bore of the Waldo Tunnel and dividing of Highway 101 would not occur for twenty more years.*

**The Marin Approach.**

*This section, reaches from the abutment to the
anchorage. Strauss: "The footings on the hillside were
built on exposed rock. Those in the valley were on
swampy ground where wooden sheet pile cofferdams
were necessary to reach suitable foundation material."
Obviously not much trouble originally, but one of the
first sites chosen for retrofit sixty years later. The
single-bore Waldo Tunnel is in the light area above the
pyramid-shaped road cut in the upper left quadrant
of this photo.*

# DARK WEDNESDAY

February 17, 1937. After concrete was poured and cured, it was necessary to remove the wooden forms from beneath the Bridge. To accomplish this, two temporary platforms were built by one of the contractors to allow workers to install and remove the wooden concrete forms.

Anyone who has seen the Golden Gate Bridge (or photos of it) immediately notices that the main span bows up at the center. The temporary platforms — sixty feet long — rolled along, hanging from the beams which made up the lower portions of the roadway framework. Because of the downward pitch from the center toward each tower, the rolling platforms had to be securely brake-clamped to ensure they would not roll away and crash into the towers.

According to most reports, there was discussion regarding the safety factor of the clamps, some saying the clamping plates and/or their bolts might be undersized for the job. What happened later proved the old adage, "No job is so important that it can excuse safety."

On this Wednesday morning, three days after Valentine's Day, 1937, eleven men worked on the platform removing wooden forms near the north tower. Two more men were in the net, cleaning up fallen debris. Stress on the west side of the platform trolley resulted in a metal failure, causing one side of the platform to fall away. Seconds later it came loose from the remaining side and the platform — all ten thousand pounds of it.

The results were disastrous. More than 2,000 feet of the net and its superstructure was ripped away and sent plunging 220 feet to the raging waters below. Twelve men went with it. Two of them were recovered alive.

The story of the accident is told in great detail in an excellent book, *The Gate*, by John Van Der Zee, published by Simon and Shuster, Inc., ISBN: 0-671-60205-5.

Strauss himself accounted for the accident succinctly in his Report of the Chief Engineer, "Concreting of the deck proceeded without serious interruption until February 17. On this date an unfortunate accident occurred which cost the lives of ten men working on a stripping scaffold near the center of the main span. In this accident, the stripping scaffold fell into the net and carried away the entire half of the main span net between the net traveler and the San Francisco Tower. Workmen were not permitted to work in the area where the net was missing."

They began to replace the net on March 3 and the job was completed April 15, 1937. However, in this one disastrous incident, Strauss' remarkable safety record would be buried under lurid, front-page headlines. Despite a flurry of investigations, none ever affixed legal blame for the accident.

**Looking Like a Real Bridge.**
*Taken in April 1937, this photo shows the three 20-foot concrete lanes have been poured, as has the east sidewalk. The west sidewalk awaits concrete and a guard rail. Opening Day is about one month away.*

## The Last Rivet.

Well, Almost. A 16-ounce rivet of solid gold from California's
historic Mother Load was donated by Mr. Charles H. Segerstrom
of Sonora. In 1865, a golden spike had been driven to mark the
completion of the first transcontinental railroad. Reportedly,
Strauss had sought to replicate a similar event for the historic
completion of the Golden Gate Bridge. The date is April 27, 1937.

**The Chief Engineer Taps Home The Last Rivet.**
*Strauss (hatless, center) readies the pound of gold for buckerup Edward Murphy (hardhat, holding tool) and riveter Ed Stanley. The big moment has arrived.*

### Let 'er Rip, Ed!

*The two workers, who have driven and bucked
thousands and thousands of real rivets, had trouble with
the unheated gold one. After it was placed, pronounced
sound by the committee, and the band music and
speeches were underway, Ed Stanley took another whack
at it. And another. Sometime later, an official noticed the
head of the rivet was gone. It was never recovered. At a
later date, another gold rivet was driven in the Golden
Gate Bridge — at a different location.*

**No Skyroom, No Restaurant, But, Yes, Elevators.**

*One of the original ideas was to have elevators for passengers to ride to the tops of the towers where they would marvel at the view from enclosed skyrooms. The budgets didn't allow realization for these amenities. However, there is a small elevator in each tower of the Bridge. This photo shows how small. With a capacity of three people — at the most — these elevators were installed to allow maintenance people access to the top of the Bridge. They are still in use today.*

### World Class Bridge Offers World Class View.

*Had there been elevators and viewing areas atop the towers, here's what tourists of 1937 would have seen. The cozy city of San Francisco nestled on its famed hills, a bustling seaport, and the Federal prison on Alcatraz (to the left of the Bridge beacon). In the background, the also-new San Francisco-Oakland Bay Bridge, completed only months before. The two bridges ended San Francisco's isolation and opened the lands east and north of the city to rapid and continuing expansion.*

**Dual Silhouettes: Man in Hat and Main Cable Saddle.**
*Details of the massive saddle, four of which carry the main cable over the tops of the towers. When warm weather expands the cables, they do not move in the saddles. Instead, the tops of the towers actually lean a few feet in each direction and are designed to move over twelve inches from side-to-side.*

## A Well-Spent $35,000,000.

*This was the sum approved by voters in 1930. The total construction cost for the Golden Gate Bridge was $27,125,000. Engineering and inspection fees were $2,050,000. Administrative fees, interest on bonds up to opening day and other costs totaled $4,068,000. Strauss declared a surplus of $1,334,000 to operate the Bridge and service the debt until the Bridge could begin paying for itself at the toll booth. This proved to be an extremely important case of fiduciary foresight.*

## Detailed Cost Breakdown:

| | |
|---|---:|
| Marin pier: | $ 436,000. |
| San Francisco Pier & Fender | 2,935,000. |
| Shore-end Pylons | 295,000. |
| Anchorages and Cable Housings | 2,328,000. |
| Tower Steel | 6,970,000. |
| Deck and Other Suspended Steel | 3,250,000. |
| Cables, Suspenders & Accessories | 5,910,000. |
| Approach Viaduct Steel & Approaches | 1,085,000. |
| Paving Bridge and Approaches | 800,000. |
| Presidio Approach Road | 1,314,000. |
| Replacement, Presidio Buildings | 575,000. |
| Sausalito Approach Road | 330,000.* |
| Toll Plaza Structures | 450,000. |
| Toll Plaza Equipment | 72,000. |
| Lighting and Electrical Installations | 270,000. |
| Tower Elevators | 60,000. |
| Miscellaneous | 45,000. |

*\*$159,000 came from the only Federal Aid to the entire project.*

# THE MIGHTY TASK IS DONE

By Joseph B. Strauss

At last the mighty task is done;
Resplendent in the western sun
The Bridge looms mountain high;
Its titan piers grip ocean floor,
It's great steel arms link shore with shore,
Its towers pierce the sky.

On its broad decks in rightful pride,
The world in swift parade shall ride,
Throughout all time to be;
Beneath, fleet ships from every port,
Vast land-locked bay, historic fort,
And dwarfing all — the sea.

To north, the Redwood Empire's gates,
To south, a happy playground waits,
In Rapturous appeal;
Here nature, free since time began,
Yields to the restless mood of man,
Accepts his bonds of steel.

Launched 'midst a thousand hope and fears,
Damned by a thousand hostile seers,
Yet ne'er its course was stayed.
But ask of those who met the foe,
Who stood alone when faith was low,
Ask them the price they paid.

Ask of the steel, each strut and wire,
As of the searching, purging fire
That marked their natal hour;
Ask of the mind, the hand, the heart,
Ask of each single stalwart part
What give it force and power.

An honored cause and nobly fought
And that which they so bravely wrought
Now glorifies their deed;
No selfish urge shall stain its life,
Nor envy, greed, intrigue, nor strife,
Nor false, ignoble greed.

High overhead its lights shall gleam,
Far, far below life's restless stream
Unceasingly shall flow;
For this was spun, its lithe fine form
To fear not war, nor time, nor storm,
For Fate had meant it so.

## The Man. The Bridge.

Joseph B. Strauss was the Chief Engineer, and through his grit and determination in handling the political and financial affairs of the early days, is the acknowledged "father" of the Golden Gate Bridge.

Unlike other engineers, however, Strauss, possessed a poet's heart, and upon completion of the Bridge in 1937, he composed this poem. It is more than a rhyme; it is an open-letter tribute to friends and foes alike. The poem alludes to the problems and the doubters with neither a mean spirit nor a boast, while saluting The Dream, the soaring artistic and structural achievement, and it concludes with the shining commitment "To fear not war, nor time, nor storm."

These words from the nineteen thirties provide a ghostly-accurate prediction of the next sixty years in the life of his beloved Bridge. Joseph Strauss died in 1938, while his creation was untroubled, shining, new-penny bright.

# OPENING DAY

## MAY 27, 1937.

### Seven Days of Celebrating

*F*rom its earliest days, San Francisco had the reputation of a town that knew how to throw a party. Spanning the Golden Gate with a structure that was a world record suspension bridge, a masterpiece of engineering, and an aesthetic and visual jewel, was plenty of cause for Mayor Angelo Rossi to plan a massive celebration.

Groups and individuals of every stripe arrived, including caravans from the eleven western states, Mexico and Guatemala, all fifty-eight California counties. Each group paraded up Market Street, where street lamps were decorated to resemble redwood trees, to the steps of City Hall where each was welcomed. The governor of Washington lead a cavalcade of seven-hundred and fifty automobiles. A choir of Mormon boys was sent from Utah. The Canadian Mounties arrived. Alaska sent dogsleds. Chile sailed one of their navy ships into the bay and marched its entire crew up Market.

Hollywood actor Robert Taylor escorted one of the many Queens chosen for the event to the coronation ball where George Jessel was the Master of Ceremonies. The whole town, it seemed, danced the night away until six a.m., Thursday, May 27 — the time of the opening of the Golden Gate Bridge.

**Pedestrian Day, May 27, 1937.**

*Unofficial, But Crowded. The first day, when vehicles weren't allowed, was not an "Official Day," when speeches and well wishing would go on. It was designated Pedestrian Day and it was for the people who had seen or heard about the mighty Bridge which had been under construction for fifty-two months. In the pre-dawn fog eighteen thousand anxious fans crowded the bridgeheads in San Francisco and Marin waiting for the Bridge to open. When it did, they crowded on to the span, excited, joyous.*

*The pedestrian toll was five-cents and was collected in boxes borrowed from the San Francisco Municipal Railway. These were jammed in early morning and for the rest of the day the toll was collected in buckets, to be counted at day's end. Still they came. By mid-morning, an estimated 45,000 had visited the Bridge. Most hadn't left. By two o'clock in the afternoon their number exceeded 100,000 and by six that evening, officials figured that as many as 200,000 people had walked onto what was now everyone's pride and joy, The Golden Gate Bridge.*

**They Came, They Saw, They Dressed.**
*With towers rising into the mist and fog obscuring much of the span, motor coaches and taxis arrive with pedestrians who want to be part of something unique and different — that "impossible" bridge they built across the Golden Gate. Although it is the fourth week of May, Northern Californians come dressed for the weather, which is often foggy and windy.*

**A Day of "Firsts."**

The first pedestrians to wave to ships from the Golden
Gate Bridge did so on Pedestrian Day. A runner from
San Francisco Junior College was the first to cross from
SF to Marin, but it was a runner from Tamalpais High
School in Mill Valley who was first to cross, Marin to
SF. And it was a runner representing the Olympic Club
who was first to cross in both directions. His time:
seventeen minutes.

Everybody tried to get into the act. Two San Francisco
sisters were the first to roller skate across. The first
priest crossed, though his name is lost. The first dog to
cross was a Scottie, and there was someone that day
who was the first to cross the Bridge on stilts. Yes, stilts.
One local newspaper began its story of the coverage
with the line, "The Old Town went nuts yesterday!"
This proved to be an understatement.

**Let The Party Begin.**
*A flight of U.S. Navy planes passed over the Golden Gate Bridge as people begin to assemble on the San Francisco side to await a motorcade that would begin in Marin. The spectators' area (center, right) would be the site of the Roundhouse, a restaurant that would be built a year later.*

**May 28, 1937** — *The Official Opening. Early this Friday morning, a group of officials of the District, governors of several states, representatives from Canada and Mexico and others, including engineers and politicians, caught a ferry boat from Fort Point to Sausalito where they formed a motor caravan and proceeded to the northern end of the Bridge. There, they found the first of three barriers installed for the ceremonies, this one in the form of two redwood logs, each three feet in diameter. Note the photographers and newsreel cameras on the stake trucks (center, left). In 1937 it was difficult to find a man (or woman) who didn't wear a hat.*

**Saw Log Sawed Same.**

*The celebration paused here while three champion
timbermen, one each from California, Washington and
Idaho, competed with handsaws for the title of
International Champion Log Sawer. The Washington
entrant, Paul Searles, won the five-hundred dollar prize
in the time of two minutes, forty-seven and two-fifths
seconds. The first barrier defeated, the officials moved
on to the next one.*

**Next Stop: The North Tower. Next Barrier: Chains.**
*The next event called for the symbolic cutting of three chains to clear the second "barrier" before the dedication. Here, Franklin P. Doyle, President of the Redwood Empire Association uses an acetylene torch to cut through a copper chain. Following this President Filmer of the Golden Gate Bridge District sliced through the links of a gold chain. Finally, Mayor Rossi of San Francisco cut through a silver chain and the Marin side of the Bridge was declared open. The motorcade proceeded across the span to San Francisco where they came face-to-face with the final barrier.*

**Here They Come!**

*This is one of the most famous photos of the Opening Day ceremonies, taken by photographers standing on those trucks seen earlier. There was an attempt to recreate this ride for the fiftieth anniversary celebration, but nothing quite compares to the statement made by this photograph, which seems to say, "We're new! we're America! We're growing, meeting challenges, looking toward the future!"*

## So Passes Power.

*Chief engineer Joseph B. Strauss (center, hat and glasses), spoke quietly to the assembled crowd of officials and well-wishers, and to much of America (via the Mutual Broadcasting System). Hands trembling, his voice barely heard above unrelenting cheers, his simple eloquence carried the day as he uttered, "This bridge needs neither praise, eulogy nor encomium. It speaks for itself. We who have labored long are grateful. What Nature rent asunder long ago, man has joined today . . ."*

## "Bridge Barrier By a Bevy of Beauties."

*That could have been the way a 1937 editor would have captioned this photo. The final of the three barriers, this one at the San Francisco end of the Bridge, was formed, not by a chorus line, but by a chain of fiesta queens, each of whom had been chosen as "Queen For a Day" during the month. This line could not be broken until Chief Engineer Strauss formally presented the now-completed span to the Golden Gate Bridge and Highway District.*

## The Golden Gate — Glory and Greatness!

By prearrangement, at high noon in his White House office (9:00a.m. in San Francisco), President Franklin D. Roosevelt pressed a telegraph key to signal the Bridge was officially open. This small electrical impulse loosened a barrage of noise such as the Bay Area had never heard. Church bells in San Francisco and Marin rang out. Foghorns throughout the Bay sounded in a unanimous, if limited symphony. Ships on the water and tied up at docks tooted. Thousand of cars, many of them lined up to cross the Bridge stood on their horns. Thirty eight ships of the U.S. Pacific Fleet began to steam under the crossing. Fireworks exploded. Champagne corks popped. The Golden Gate Bridge was open!

**There Was Quiet, Too.**

*Not everyone went delirious. Some stood in silent wonder at the great bridge and the views it offered. This photo, more than most, reveals the Art Deco design details. The San Francisco Tower (right) also reveals why it took more than 600,000 rivets to assemble the big orange bridge across the Golden Gate.*

**And Then There Were None.**

*Something of a disappointment, and perhaps a sign of things to come, vehicular traffic on Opening Day was far fewer than expected — only 32,300. Based on the overwhelming success of Pedestrian Day twenty-four hours earlier, some officials surmised that people may have felt they could drive across the Bridge anytime, but there would be only one chance to walk across in complete freedom.*

**The Soft Glow of the Future.**

*Despite the celebration and accolades from around the world, the Golden Gate Bridge did not exactly face a golden future in May 1937. There were still rumors that it was unsafe, despite the fact 200,000 people had jammed onto it on Pedestrian Day. Non supporters also promoted unfounded alarm that a slight earthquake would dash it into the Bay. There was even a fable about how a lone violinist could walk onto the Bridge, and with his bow strike a particular note that would result in the harmonic destruction of the Bridge, shore to shore. None of these was true and on a soft night in May, the Golden Gate Bridge's glow seemed to emphasize that the political and engineering challenges had been met. What is not reflected in this graceful portrait is the looming challenge it now faced: Succeeding as a business enterprise.*

# THE BRIDGE IS BUILT

*[We continue now with the District's first Secretary, W.W. Felt, Jr., and his story of the early days of Bridge operation. From this 1940 pamphlet, he summarizes the construction of the Bridge (as we have just seen) in a few paragraphs, and manages to put all of the problems succinctly in the background. After all, the Bridge is completed now and Felt's prose does not dwell on the past, but on the present.]*

The work after starting on January 5th, 1933, proceeded without strike, or any unusual incident, excepting that soon after the completion of the access trestle from the shore to the San Francisco pier site, a vessel off its course in a thick fog, crashed through it, causing considerable damage. The trestle was repaired, and almost immediately a storm carried away about 800 feet of it. The trestle was then raised and anchored to bedrock, and it thereafter remained in place until removed after completion of the Bridge. An outstanding feature in the construction of this great structure was the protection of the workmen from accident. The latest and most up-to-date safety measures were in use to guard the men, and included a safety net suspended beneath the structure. This net was made of 3/8" diameter Manila rope, with 6" centers both ways. The use of this net was responsible for saving the lives of 19 men. From the beginning of the work up to February 17th, 1937, there had been but a single fatality, and the work bid fair to be completed with the loss of but a single man, but on February 17th, the unexpected happened when a stripping platform suspended beneath the bridge, upon which 12 men were working, gave way and plunged into the waters of the ocean, carrying down with it these men and approximately 2000 feet of the safety net. Only two of the men were recovered alive. The net was replaced and the work proceeded without further delay or loss, to completion.

The work of completing the world's longest single span suspension bridge was now brought to a close and on May 27th, 1937, the Bridge was thrown open to pedestrians exclusively, for that day, and approximately 200,000 persons took advantage of the opportunity, and enjoyed the thrill of walking across the Golden Gate. On the following day at high noon, attended by appropriate ceremony, the bridge was officially opened to vehicular traffic.

Thus were the hope of that courageous pioneer group fulfilled and the challenge met for all time.

# BRIDGE OPERATIONS

*T*he operation and maintenance of the bridge is directed from the Administration Building situated at the Toll Plaza.

The position of the toll collector is a very important one, for the toll collector is the contact man of the Bridge District with the public, and he is required at all times to maintain a friendly, courteous attitude toward patrons of the Bridge.

The collection of tolls is performed by a squad of sixteen full-time collectors ably directed by a toll captain with the assistance of four sergeants. Additional part-time men are used, as traffic conditions require.

Fourteen lanes are provided, through which traffic is passed for the purpose of toll collection. These lanes may be opened or closed according to traffic requirement. They may also be opened to accommodate traffic from either direction.

Communications between the Toll Captain, the Sergeant and collections on duty is by means of "Voycall", a two-way system permitting direct conversation between a collector and his supervising superiors. In addition, the Captain or Sergeant may, by opening a switch, listen in on a conversation in any lane between the collector and a patron. The advantages of this feature are many and obvious.

*From a brochure written by Golden Gate Bridge and*
*Highway District Secretary, W. W. Felt, Jr., circa January 1940.*

# RECORDING TOLLS

The toll recording system is said to be the fastest electric recording in use on any bridge, the time required for an average toll transaction not exceeding six seconds. The manufacturers' recording capacity rating for this equipment is 600 cars per lane per hour. This number, however, has been exceeded when 647 cars passed through a single lane within one hour.

Toll collections are recorded on a separate recording unit for each traffic lane. These recording units (14 in number) are housed in the recording room, and are under the direct control of the auditor.

Upon going on duty, the collector inserts his identification key into the registry box of the particular lane for which he is collecting. This operation unlocks the registry equipment, turns on the electric current for that unit, and registers his identification number on the recording tape.

A collection, upon receipt of a toll, punches a button in the registry box, which operation records the amount paid in the recording unit, and indicates the amount in a cabinet in the lane; the vehicle classification is also indicated by number on the overhead indicator for the lane. Thus the patron and those on duty in the Toll Captain's office may observe the amount of toll recorded and from other observation points, the classification as shown on the overhead indicators may be checked against the vehicle passing through the traffic lane.

Upon leaving the collection point, the vehicle passes over a treadle which depresses, and also automatically records on the axle counter, the number of axles on the vehicle. At the end of each hour, total of transactions of corresponding traffic lanes are automatically recorded on the recording tape of each unit.

*From a brochure written by Golden Gate Bridge and*
*Highway District Secretary, W. W. Felt, Jr., circa January 1940.*

# COUNTING MONEY

Upon completing his shift, a collector removes his identification key, thereby locking the recording mechanism for his particular lane. He then makes up his cash, placing each coin denomination in a separate bag. His report is made in duplicate, and the receipts, together with a copy of the collector's report, are all placed in a large sealed bag, which, in turn, is deposited through a chute into the money room situated below the roadway level.

The teller counts the money and checks against the collector's report. If any mistake is found, the error is noted on the report, which is then transmitted to the auditor, and checked against the automatic recording record.

The teller then bags the coins by denominations, and in amounts as prescribed by the District's depository bank. All coin counting is performed by automatic coin counting machines. The currency is segmented into denominational packages, banded and, together with bagged money, deposited through a non-returnable deposit head into a burglar-proof vault. This vault bears a time-lock combination which is in the possession of the depository bank. At frequent intervals, the money is removed by the bank and transported by armored cars for deposit.

# MAINTENANCE

The maintenance work on the bridge and its approaches is performed by men well equipped for that type of work, and they receive the regular scale of wages in the area for work of similar character.

Electricity plays a prominent part in the bridge operation, and a chief electrician, together with a number of capable assistants, is in active control of the electrical equipment at all times.

A steel structure, in such an exposed position as the Golden Gate Bridge, must, at all times, be kept well painted. Therefore, skilled painters in number as required, give careful attention to this important item of maintenance.

The quality of paint used was determined after three years of paint testing. These paints were submitted (upon invitation of the chief engineer) by various paint manufacturers, and in selecting the color, durability, as well as the artistic value, was given due consideration. As regards the latter, advice was had from a number of well-known artists and color authorities; hence the color "International Orange" was chosen, not only for its durability, but for its pleasing effect and harmonizing quality with the surrounding landscape.

*From a brochure written by Golden Gate Bridge and*
*Highway District Secretary, W. W. Felt, Jr., circa January 1940.*

## SAFETY & PATROL

*A* tow car and a fire truck are stationed at the toll plaza, each on twenty-four hour service, and operated by competent men.

The policing of the bridge and approaches is performed by a squad from the California Highway Patrol, consisting of eleven officers under the direction of a captain and two sergeants.

That the policing is done effectively, is attested by the fact that in thirty-two months of operation, only one accident has occurred on the bridge or its approaches resulting is loss of life or serious injury to person or property. Fifty-one thousand, seven hundred and eighty-five vehicles crossing the bridge in a single day without an accident of any kind, bears splendid testimony to the effective work of this small group of intelligent and highly trained men.

## VISITOR AMENITIES

*T* elephones are located at convenient points on the bridge, from which a patrolman, the towing car or the fire apparatus may be called.

Parking space is conveniently located at the toll plaza for those who may wish to stop their cars for the purpose of walking on or across the bridge.

Dinner may be enjoyed at a restaurant conveniently situated near the parking area. The restaurant building is circular in form, and permits panoramic view of the surrounding country. This view includes the Golden Gate Bridge, the Marin hills, the harbor itself, with the cities of Richmond, Berkeley, Oakland and Alameda forming a background. Grim Alcatraz is seen seemingly but a stone's throw away; Angel Island, quiet and peaceful, Treasure Island with its Golden Gate International Exposition opening early in 1939 and to continue during a portion of 1940 fairly radiating enchantment and beauty from this magic isle created by the hand of man — a fitting commemoration to the erection of two magnificent bridges; San Francisco, from its many hills, proudly proclaiming to the world the courage and daring of its people in overcoming whatever may claim to impede its progress.

*From a brochure written by Golden Gate Bridge and
Highway District Secretary, W. W. Felt, Jr., circa January 1940.*

# DETAILS, DETAILS

The District has provided, maintains and operates, under government requirement, a lighthouse on the San Francisco pier; revolving air beacons atop each main tower; pier lamps; mid-channel lights; obstruction lights on each cable; two diaphones at the center of the main span, and two typhones at the San Francisco pier. [Note: Diaphones and typhones are better known by their simpler name: Foghorns.]

It is said that the lighthouse is the only one to be situated on a bridge, and that the diaphones are the only fog signals in the world toward which boats may steer, instead of avoid. [This is still true and the foghorns are still used. Despite the use of radar on modern vessels, San Francisco Harbor is home to thousands of small yachts and sailing craft. Mariners entering the Bay for the first time in foggy conditions admit their wariness at aiming toward a foghorn instead of away from it. When in use, the mid-span foghorn can be heard across many parts of San Francisco and in the small cities of southern Marin county.]

The navigation lights and signals must be ready to function at all times, and in addition to the regular equipment and power source, emergency stand-by service is instantly available, if required.

The roadway throughout the entire length of the main structure and the approaches is lighted by sodium vapor roadway lights, mounted on specially designed electroliers, and spaced 150 feet apart on either side of the roadway.

A small electric elevator is installed in the east leg of each of the main towers, and is used for maintenance and inspection purposes and in gaining access to the tower base, the top of the tower, and for each intermediate strut.

*From a brochure written by Golden Gate Bridge and*
*Highway District Secretary, W. W. Felt, Jr., circa January 1940.*

# BENEFITS

*T*he construction of the Bridge closed the gap in California-U.S. Highway 101, the principal route along the coast for the entire length of the state. This important and scenic road is an integral part of the great international highway eventually to stretch the length of the North and South American continents.

As the people of the two continents intermingle by reason of travel over this international highway, a better knowledge and understanding of each other will naturally result, and promote the "good neighbor" policy between the United States and her sister countries, constituting in effect, a good-will ambassadorship as nothing else may.

The importance of the Golden Gate Bridge as part of our military defense is apparent when we consider that it directly connects the Presidio at San Francisco with Ft. Baker military reservation in Marin County, and, in addition, is a means of connecting Fort Funston, Sunnyvale Air Base, Monterey Presidio, Hamilton Field and the Mare Island Navy Shipyard.

By means of the Bridge, the equipment and personnel stationed at these points may be mobilized in approximately minutes as the hours heretofore required when transportation was by boat.

Even as the building of its bridges made possible the great metropolitan area of New York, so may San Francisco, with its bridges (present and future) develop an extensive metropolitan territory of approximately one hundred miles square with boundaries extending to Santa Rosa in the north, Stockton in the east and San Jose in the south.

*From a brochure written by Golden Gate Bridge and*
*Highway District Secretary, W. W. Felt, Jr., circa January 1940.*

# CASH SAVINGS

uring the time the Bridge has been in operation, there has been saved to the motorists using the Bridge approximately $4,849,763.00. This sum represents the difference between what the motorists actually paid and what they would have had to pay under the old regime — being due to the reduction of tolls brought about by the construction of the Bridge.

During the same period, 389,803 Federal Government vehicles and vehicles used by army and navy personnel crossed the bridge without payment of tolls. This government traffic, if crossing by boat and paying the toll which would have been required to be paid to the Ferry Company would have amounted to a sum in excess of $200,000.00.

With the estimate number of persons in vehicles crossing the Bridge during the 32 months' period figured at 29,168,026, the time saved by the Bridge over the slower ferry system would have amounted to approximately 9,722,675 man hours, which if valued at 25 cents per hour would amount to a further saving of $2,430,668.75. This saving in time and money furnishes concrete proof of the further savings which may accrue to motorists traveling the Golden Gate Bridge.

Long may this Bridge stand, a monument to man's achievement, a symbol of strength and beauty, and may it ever continue to fulfill the purpose for which it was intended.

*From a brochure written by Golden Gate Bridge and*
*Highway District Secretary, W. W. Felt, Jr., circa January 1940.*

# From Then Until Now...
# The Bridge Writes
# Its Own History

*T*his is where most books about the Golden Gate Bridge end. However, the history of this great Bridge since Opening Day in 1937 is every bit as exciting as the events which lead to its construction. At times the story offers the high drama of a suspense novel. At others, it is a primer on the politics of California. In between, there is the continuing struggle of the Bridge to change with the times, to explore their core mission, to lead in innovative traffic management and toll collection methods.

In the six decades since completion, the Golden Gate Bridge has been tested by war, by time and by storm, it has also experienced challenges no dreamer-engineer ever imagined. It is those years and those challenges to which we next turn our attention.

Here, for the first time, is what has happened to this now-historic structure since its completion, told in the words of the people who have been responsible for the Bridge, responsible for keeping The Dream alive, through good times and perilous times, through hope and through hardship. As you'll see, the Golden Gate Bridge adventure did not end on Opening Day.

That was only The Beginning.

SIXTY YEARS OF SERVICE

# Highlights from Annual Reports

One of the best ways to understand an institution is to examine the documents created from its birth up to present day. A study of the Annual Reports of The Golden Gate Bridge, Highway and Transportation District, as prepared by the Staff and Board of Directors though its history, reveals the care and concern with which the affairs of the Bridge were managed from its first beginnings.

Fiscal responsibility, safety, maintenance and consideration of future needs were issues addressed in the very first Annual Report and they remain primary concerns today.

Early concerns also reflect the Board's understanding of the importance of the Bridge as a working partner in both the commerce and recreation of the day. Great emphasis is put on improving approaches to the Bridge, both from San Francisco and on the Marin side, especially the Sausalito exit and state roads to the north.

Later, as new problems arise, new challenges face the Bridge, the Board, its Officers and its Employees. As each challenge appears, it is met with courage and responsibility, and with a sense of duty to the communities and public it serves.

These highlights are edited for space and conform to the originals. Bridge scholars and others may note the use of language, capitalization and punctuation changes over the years, as do the names of many of the agencies with whom the Golden Gate Bridge, Highway and Transportation District does business.

Sometimes, "the Bridge" becomes "the bridge," without a capital letter. The same thing happens to the Board, the Directors, the General Manager. Oddly, "the Toll Plaza" becomes "the toll plaza" only a few times.

This type of inconsistency is not peculiar to the Annual Reports of the Golden Gate Bridge, Highway and Transportation District, but is comparable to the floating stylesheets common to all companies through the years.

Please note, as we have covered the years of construction, leading up to the opening of the Bridge, our Historical Snapshot briefs are for the second-portion of each Fiscal Year.

SIXTY YEARS OF SERVICE

# Highlights from Annual Reports

YEAR: 1937 — 1938

At the end of the first full fiscal year on June 30, 1938, the Bridge had served 3,892,063 motor vehicles, carrying more than 8,000,000 passengers, and in excess of 400,000 pedestrians had walked upon its two 11-foot sidewalks, over 250 feet above the waters of the Golden Gate. *[Note: This was a slight exaggeration; the Bridge deck is actually 220 feet above the water.]*

These 3,892,063 motorists paid an average of 48.5 cents, resulting in revenues of $1,907,776.57 to the Golden Gate Bridge and Highway District, but at the same time representing a saving under previous ferry tolls of over $2,000,000.00. These motorists, furthermore, each completed the North Bay crossing with a considerable saving of time which, if reckoned as a whole, could easily be calculated as an additional community saving of $1,500,000 to $2,000,000 or more. In these figures may be found a complete justification for the construction of a $35,000,000 bridge across the Golden Gate of San Francisco Bay.

A complete record of traffic, by classifications, is included in this report. This record reveals the highly seasonal character of the traffic, it being highest in the summer months when the numerous popular resorts of the Redwood Empire, especially the Russian River region, are open and lowest in the winter months. Sundays and holidays in the summer are the extreme peak days, although there has been a tendency for the differential to lessen, due in large part to the extreme congestion formerly encountered in San Rafael and northerly along the reconstruction area. There has also been a tendency for resort patrons to return later on Sunday night, many returning only in time for work on Monday morning. This reflects a new freedom from boat schedules and a considerable lengthening of the weekend recreation period. The necessity for being prepared to handle peak traffic loads increases somewhat the expense of operating, but it is a factor which cannot be controlled.

The volume of free traffic of the United States Government, i.e., Army and Navy vehicles and vehicles of Army and Navy personnel (including retired personnel) and their dependents, has been far beyond all expectations. This traffic is handled by the District at considerable extra expense in spite of the fact that the great majority of the traffic is not official Army or Navy business.

## Additions and Betterments

The District has constructed and opened a parking area of adequate size to accommodate motorists desiring to stop at the Toll Plaza for a walk on the Bridge. Generous use has been made of the facility since it was opened in January, 1938.

The need for more adequate restroom facilities for pedestrians and the lack of an accessible eating place for employees and pedestrians has been met by the construction, at the Toll Plaza, of a building to house a restaurant and rest-rooms. The building is circular in plan and affords an unexcelled panoramic view of San Francisco, the Bay and the bridges. The restaurant was opened on May 27, 1938 and is enjoyed increasingly by tourists, commuters, pedestrians and San Franciscans.

A small parking area and a facilities building have also been provided at the northerly end of the Bridge. The building houses the pedestrian turnstile and restrooms and has a limited space for a concession.

---

*In the rest of the World . . . .*

## HISTORICAL SNAPSHOT

*The Bridge was conceived and built during an important, expanding time in history. It's birth was difficult; it exists in a real world. These are a few of the events, near and far, that happened concurrently with its development and growth. As you'll see, from time to time events occuring far from San Francisco Bay held influence over the business of the Golden Gate Bridge.*

**Events.** *In U.S. minimum wage established, 40 cents per hour. Howard Huges flies around the world in 3 days, 19 hours, 17 minutes. Luxury Liner, Queen Elizabeth, launched. Introduction of fibreglass and teflon.*

**Literature.** *Someset Maugham, The Summing Up; C. Isherwood, Goodbye to Berlin; e e cummings, Collected Poems; Daphne du Maurier, Rebecca.*

**Drama.** *Jean Cocteau, Les Parents terribles; Thornton Wilder, Our Town.*

**Music.** *Paul Hindemith, Mathis der Maler (opera); Sergei Prokofiev, Romeo and Juliet (ballet); Irving Berling, "God Bless America;" Benny Goodman concert at Carnegie Hall*

**Sports.** *At World Cup in Paris, Italy defeats Hungary 4-2. NY beat Chicago Cubs 4 straight in World Series.*

**Films.** *Snow White and the Seven Dwarfs; Olympiad, director Leni Riefenstahl; The Adventures of Robin Hood, starring Errol Flynn; Le Quai des Brumes, director, Marcel Carne.*

**Radio.** *Orson Welles' CBS broadcast, "War of the Worlds" startles U.S.*

**The Golden Gate Bridge had completed Year One.**

# 1938

## In Memoriam

*MAY 16, 1938*
### JOSEPH B. STRAUSS

**CONSTRUCTION OF THE
GOLDEN GATE BRIDGE
CULMINATED HIS GREAT
CAREER AS A BRIDGE
ENGINEER**

*[Editor's Note: This is why the main street through Sausalito is called "Bridgeway." The City of Sausalito fought this proposal and ultimately improved truck design and more powerful engines eliminated the need for this lateral or "bypass."]*

### Safety Record

An enviable record in safe driving was established on the Bridge and its approaches during the thirteen months and four days ended June 30, 1938. Only six accidents occurred during this period while 3,892,063 motor vehicles were traversing the 4½ miles of Bridge and approach roadways. There were no fatalities and no injuries requiring hospitalization. Although every precaution was taken in the construction of the Bridge to lessen the accident hazard, a large share of this fine record is due to the tireless efforts of the Golden Gate Bridge Detail of the California Highway Patrol. This fine group of twelve men has guarded the safety record zealously since the day the Bridge opened.

### Highways and Approaches

The District has been seriously handicapped in developing revenue traffic and motorists have been greatly inconvenienced by the lack of adequate approaches to the Bridge and by the condition of highways tributary to the Bridge. The improvements needed are:

1. Construction of the proposed Funston Avenue Approach through the Presidio.

2. Widening of Lombard Street.

3. Realignment and reconstruction of the highway northerly from San Rafael to Santa Rosa.

4. Extension of the Sausalito Lateral.

5. A direct connection between the Sacramento and San Joaquin valleys and the Redwood Highway at Ignacio.

The commencement of construction on the Funston Avenue Approach, after meeting with numerous obstacles, appears to be scheduled for December, 1938 . . . and may require two years to complete.

The greatly increased volume of traffic on Lombard Street (San Francisco), particularly heavily laden truck traffic, has brought about a condition that requires immediate correction. The necessity for widening Lombard Street has been acknowledged for several years and now the street surfacing is badly in need of extensive repairs. It is hoped that the City of San Francisco will commence this widening project in the early future.

The consideration of Lombard Street is closely associated with the projected construction of vehicular tunnels on Lombard, Broadway and Divisadero Streets. A Lombard Street tunnel would provide a direct connection from the Bridge with Columbus Avenue and The Embarcadero, relieving traffic congestion on Bay Street and eliminating a difficult grade for trucks. It would also provide a fast route for commuters and others destined for the downtown district. The District is very much interested in these proposed tunnels.

The State Highway Department has already taken definite steps to improve conditions in San Rafael and northerly to Ignacio, the present construction program being scheduled for completion by October 1, 1938. The opening of this improved section will greatly benefit the Bridge but there will still remain a serious need for highway realignment and widening northerly through Petaluma and beyond to the Russian River resort country, destination of thousands of peak-period motorists.

The continuous 6% grade of the Waldo approach road, built by the State Highway Department imposes a heavy burden on truck operators. This burden would be materially decreased by an extension of the Sausalito Lateral. The summit on the Waldo approach is approximately 366 feet above the north end of the Bridge and it would be possible to eliminate this additional climb if the Sausalito Lateral were extended to water level through the City of Sausalito. (See note above, left.)

The District continues to be concerned with obtaining a suitable highway connection to the Sacramento and San Joaquin valleys. The State Highway Department has indicated a willingness to purchase and improve the Sears Point Toll Road but up to this time negotiations have not been completed. □

SIXTY YEARS OF SERVICE

# Highlights from Annual Reports

YEAR: 1938 — 1939

The second fiscal year of operation of the Golden Gate Bridge has been concluded, the 22% increase in traffic over the first year proving beyond any question of doubt the economic value of the world's greatest suspension bridge.

A total of 4,031,504 vehicles crossed the Bridge in 1938-39, compared with 3,311,512 in 1937-38, an increase of 719,992 vehicles.

The legislative act establishing the District provides that tolls be collected sufficient to pay all obligations, including operating expenses, repairs and depreciation, interest on bonded indebtedness and, so far as possible, provide for the retirement of bonds at maturity. As yet the volume of traffic has not reached a level sufficient to assure such returns, and although it is the desire of the District to stimulate traffic in every way possible, a reduction in tolls cannot be resorted to for this purpose until it can be shown that such a reduction will result in no diminution of total revenue below the required amount. In this connection the District has observed with interest the results of toll reductions on other bridges within the past two years, and in no case has it been found that a reduction in passenger automobile toll has resulted in an increase in revenue. On the contrary, a definite revenue loss has resulted from such reductions.

The Northwestern Pacific Railroad Company, which has for a number of years operated a combined train and ferry service between San Francisco and Marin County points, petitioned the State Railroad Commission early in 1939 for permission to abandon its services because of heavy financial losses. Ferries of this company have been engaged chiefly in transporting foot passengers, although limited vehicle-carrying and freight services have been offered. At the end of June 1939 hearings on abandonment were continuing before the Railroad Commission, which body also had under consideration the application of Pacific Greyhound Lines, Inc. to serve the Marin County area with a

system of bus transportation. In order to meet this situation and to make it possible for the community to obtain an adequate system of mass transportation, the District established special rates for a bus company operation such a service. Instead of the regular 75 cents for bus and driver plus five cents for each passenger, the qualifying company would pay 50 cents for each bus crossing plus five cents for each non-commutation passenger and two and one-half cents for each commutation passenger.

It is estimated that should Pacific Greyhound Lines operate this service, bus traffic across the Bridge will be increased approximately 400%, while net bus revenue may be increased by approximately 100%.

---

*In the rest of the World . . . .*

## HISTORICAL SNAPSHOT

*The Bridge was conceived and built during an important, expanding time in history. It's birth was difficult; it exists in a real world. These are a few of the events, near and far, that happened concurrently with its development and growth. As you'll see, from time to time events occuring far from San Francisco Bay held influence over the business of the Golden Gate Bridge.*

**Events.** *In Spanish Civil War, Madrid surrenders, war ends. Hewlett-Packard company founded in California. New York World's Fair opens. A. Einstein writes to President Roosevelt about possibility of creating atomic power. FM radio developed. PAN AM begins transatlantic service.*

**Literature.** *Raymond Chandler, The Big Sleep; John Steinbeck, The Grapes of Wrath; Thomas Mann, Lotte In Veimar; J R R Tolkien, The Hobbit; Jan Struther, Mrs. Miniver.*

**Drama.** *Philip Barry, The Philadelphia Story; Lillian Hellman, The Little Foxes.*

**Music.** *Bartok, String Quartet No. 6; Marian Anderson sings at the White House.*

**Sports.** *Lou Gehrig retires.*

**Films.** *Gone With the Wind; The Wizard of Oz; Ninotchka, starring Greta Garbo.*

**Television.** *NBC begins first regular service.*

**The Golden Gate Bridge was becoming popular . . .**

# 1939

## Factors Contributing To Traffic Increase

On July 25, 1938, the Southern Pacific—Golden Gate Ferries, Ltd. suspended operations between San Francisco (Hyde Street) and Sausalito, as a result of which traffic on the Bridge was immediately stimulated. In every month after July there was an increase in traffic over the corresponding month of the preceding year.

The action of the State of California in eliminating tolls on the Sears Point road (Sacramento route), in completing the San Rafael—Ignacio highway realignment and widening and in contributing to the purchase of the Muir Woods toll road has resulted in definite increases in traffic across the Bridge.

Since February 1939 there has been the further factor of Exposition travel, but it is impossible to determine the effect on Bridge traffic. *[Note: This is the U.S. West Coast International Exposition. There was also a World's Fair in New York the same year. — Ed.]* It appears evident that the Golden Gate International Exposition has favorably affected week-day traffic, because of the many thousands of cars brought to the Bay region; however, as summer approached, it also became evident that the Exposition might be keeping many motorists from taking their accustomed week-end trips into Marin County and the Russian River region, as Sunday and holiday traffic failed to show corresponding increases over the previous year.

## Maintenance

To safeguard the use of the Bridge without interruption for heavy repairs, the management deems it necessary to keep the property of the District — the Bridge, its approaches, buildings and equipment — in full repair. Maintenance of the structure, buildings and equipment required $170,582.97 during the past fiscal year.

**Painting.** The largest item in the maintenance budget is for painting the structural steel. During the past fiscal year the sum of $112,431.84 was expended for painting the Bridge structures. This amount was divided as follows: labor, $86,589.13; brushes, tools, etc. $5,430.67; paint, $20,412.64.

The exposure to salt-laden fog is more severe at the Golden Gate than on any other bridge in the Bay area. Not only is the fog extremely active in attacking the paint film but it also limits the hours when painting can be done. Over thirty per cent of the working hours during the last fiscal year could not be utilized for outside painting because of weather conditions.

The condition of the paint on the steelwork was so critical at the time of opening of the Bridge that an acceleration of the original painting program was necessary. This situation arose because two conditions were not thoroughly understood at the time the original estimate of operating costs was made, namely: (1) the severity of the exposure and (2) the importance of special treatment of the steel surfaces prior to erection. On steel structures erected subsequent to the Golden Gate Bridge, steel surfaces have been treated by sandblasting or flame cleaning prior to erection. Failure to use either of these special treatments has added greatly to the cost of maintenance painting. The six months' delay in the construction of the Bridge also contributed to this additional cost since, at the time the Bridge was opened, the main towers, or over forty per cent of the tonnage had already had an exposure of one year. Prior to completion of the Bridge, the contractors and the District found it necessary to expend over $130,000 for paint maintenance on the towers and the main span.

Coincidental with painting operations, the District is continually experimenting with new paint materials as they are developed and offered. Many full-size panels of the Bridge are covered with these experimental samples and their resistance under actual exposure on the Bridge is being closely watched and noted.

During the past fiscal year the exterior of the San Francisco Tower has been completely repainted and the interiors of both towers have been spotted where most needed. Painting of the stiffening trusses and floor steel has progressed outward from both towers and approximately twenty-eight per cent of this tonnage has been covered.

## Highways and Approaches

The financial success of the Bridge District depends almost entirely upon the adequacy of approach roads and streets. The volume of traffic required to meet District expenses and make it possible to reduce tolls can only be developed if approaches are adequate to handle the increased traffic. This factor has been one of the District's greatest handicaps and while some progress is being made, a serious condition still exists.

As an indication of the progress being made, the City and County of San Francisco has completed a parkway boulevard between the Presidio and Golden Gate Park, is rapidly completing a divided boulevard through the park to Nineteenth Avenue and is progressing satisfactorily with the widening of the latter into a six-lane divided thoroughfare. These projects, when completed, will provide a splendid connection with the Funston Avenue approach to the Bridge through the Presidio. This project is now under construction by the State Highway Department with City and County of San Francisco gasoline tax funds, State Highway revenues and a W.P.A. grant. It is scheduled for completion by April 1, 1940 and will provide an important approach serving the Richmond, Sunset, Parkside and Twin Peaks residential districts of San Francisco as well as through traffic to the south.

It was stated in last year's report that the condition of Lombard Street required immediate attention. No apparent progress has been made toward the widening and improvement of this important street and the increasing volume of traffic, particularly heavily laden trucks, has created a serious condition. Unless the widening program can be commenced shortly, extensive costly repairs must be made by the City.

Increasing Bridge traffic has augmented the congestion of the two-lane highway between Ignacio and Santa Rosa (and beyond). The State Highway Department plans to eliminate the dangerous bottleneck in Petaluma during 1940.

Efforts were continued and redoubled during the year to obtain a satisfactory connection through the City of Sausalito Lateral to the Bridge. This improvement is seriously needed to provide a low-level approach for trucks and buses.

The management of the District appreciates the efforts that have been made by the State Highway Department to provide adequate approaches to the Bridge. Traffic over roads leading to the Bridge has more than doubled since the opening of the Bridge, and this unprecedented condition has created serious problems for the State Highway Department. Roads that were already inadequate to handle the former traffic load have now become hopelessly overcrowded. Nevertheless, we feel the State Highway Department is making a real effort to modernize these roads with the funds available. Similar problems have confronted City and County of San Francisco officials, and here again it should be reported that efforts are being made to accommodate the increased volume of traffic.

## Safety Record

The $4\frac{1}{2}$ miles of Bridge and approach roadways constructed by the Golden Gate Bridge and Highway District continued during 1938-39 to be one of the world's safest thoroughfares. In the first two years, one month and four days of operation, only fourteen accidents occurred. There were no fatalities and no injuries requiring hospitalization. All accidents were minor and vehicles involved were removed under their own power. This is a most enviable record in safe driving, particularly in view of the 35,000,000 vehicle-miles driven on the Bridge and approach roadways during this period, and reflects the precautions that were taken in the construction of the project to minimize accident hazards, as well as the very efficient work of the Bridge Detail of the California Highway Patrol. This unit of the Highway Patrol has established a record that will be hard to equal. □

**Seals Stadium.** *Baseball fans from northern counties crossed the Golden Gate Bridge to visit San Francisco's Pacific Coast League baseball team at the site of its games in an industrial section of the city. The park was located adjacent to a brewery and remained the center of baseball until the New York Giants moved to San Francisco in 1958. The new team played here until completion of its stadium at Candlestick Point.*

## SIXTY YEARS OF SERVICE
# Highlights from Annual Reports
## YEAR: 1939 — 1940

The operating results of the Golden Gate Bridge in the third fiscal year, ended June 30, 1940, reflect the very favorable influence of reduced competition and improved approach roads.

The number of toll-paying vehicles increased 6% over the previous year, being 4,098,707 compared with 3,866,959, an increase of 231,748 vehicle crossings. In addition, there were 217,160 non-revenue vehicles in the Army and Navy classifications, an increase of 164,545 such vehicles in the previous year.

Another significant characteristic of the 1939-40 traffic was an increase in bus passengers and an 11% increase in bus crossings. Compared with the first fiscal year (1937-38) bus passengers have increased 79.9% and bus crossings 34.5%. These substantial increases have occurred despite the fact no bus commutation service has as yet been inaugurated and in the face of continued ferry passenger service competition.

### Maintenance

Total expenditures in the fiscal year for current repairs and maintenance amounted to $175,047.30.

The largest item in the maintenance budget for the past fiscal year was for painting which cost the District the sum of $119,335.65. Repainting and cleaning the steel of rust and millscale was confined solely to those surfaces where the original paint coat protection had been weathered away or destroyed by millscale lifting. No surfaces already painted by the District's force have yet required a second treatment. Of the original paint coats yet to be treated there remains the Marin Tower, main cables, 32% of the suspended structure and most of the approach steel.

The sum of $100,000 has been budgeted for painting in 1940-41.

Regarding Highways and Approaches During the fiscal year 1939-40, very satisfactory progress has been made.

The acquisition of the property necessary for the widening of Lombard Street between Richardson Avenue and Van Ness Avenue has progressed to a point where a large part of the properties necessary to the widening has been purchased and some of the improvements on these particular properties have been set back thirty feet from the present southerly line of Lombard Street.

The highway between San Rafael and Ignacio has been widened so as to provide four lanes in the greater part of this distance. There are some stretches, however, where only three lanes have been provided. The highway between San Rafael and Richardson Bay viaduct is now being improved to provide four lanes of traffic over this entire distance. The widening of the highway immediately adjoining Petaluma is now being done so as to provide four lanes both northerly and southerly therefrom.

---

*In the rest of the World . . . .*
## HISTORICAL SNAPSHOT

**Events.** *FDR re-elected president of U.S. Grant to build giant cyclotron at UC Berkeley. Penicillin developed in England. First successful helicopter flight in U.S. by Vought-Sikorsky Corp.*

**Literature.** *Hemingway, For Whom The Bell Tolls; Arthur Koestler, Darkness at Noon.*

**Born.** *Jack Nicklaus, John Lennon, Pele.*

**Deaths.** *Leon Trotsky, F. Scott Fitzgerald.*

**Music.** *Prokofiev, Piano Sonata No. 6; Webern, Variations for Orchestra; Stan Kenton band founded.*

**Films.** *Hatti McDaniel first African-American woman wins Supporting Role Oscar; Fantasia, produced by Disney; First "Road" pictures (Singapore) starring Bing Crosby, Bob Hope, Dorothy Lamour.*

**Everyday Life.** *Nylon stockings introduced in U.S.; Willys Corp introduces the Jeep. Col Saunders founds KFC.*

# 1940

## An Appreciation

The Directors and Officers of Golden Gate Bridge and Highway District congratulate

### Frank P. Doyle

upon his completion of fifty years of service to the Exchange Bank of Santa Rosa, to the County of Sonoma and to the Redwood Empire, and record with particular pride his constant devotion to the task of promoting and constructing the Golden Gate Bridge. His valuable services in this great undertaking have continued tirelessly since the very inception of the project. It has been the special privilege and pleasure of the undersigned to call him friend and colleague, and it is fervently hoped this association may continue for many years.

San Francisco, May 1, 1940

*This congratulatory scroll was awarded to Frank P. Doyle in 1940. It was Doyle, a Santa Rosa banker and then president of the Chamber of Commerce, who called a meeting in support of the idea of a Golden Gate bridge on January 13, 1923. On Opening Day, he cut one of the chains to officially open the Bridge. The southern approach to the Golden Gate Bridge was named Doyle Drive in his honor.*

A contract has been let by the State Highway Department and construction has begun on relocation and improvement of the highway through the City of San Rafael. A viaduct is being constructed to separate grade crossings at five different street intersections. This improvement has been necessary due to the extreme congestion resulting at times of peak traffic flow.

The Funston Avenue Approach to the Bridge through the Presidio of San Francisco was completed during the fiscal year just closed, and is now open to traffic. This improvement is serving the purpose for which it was intended. There is a large amount of traffic over this approach and that portion of our ramp between the cloverleaf connection and the Marina.

There is, however, one project which is highly desirable on which little progress has been made, the importance of which is emphasized by the traffic congestion which has occurred on the Waldo Approach to the Bridge. This project is the extension of our Sausalito Lateral through the City limits of Sausalito in order to provide a low-grade approach to the Bridge, avoiding the long 6% grade on the Waldo Approach.

This matter is now in the hands of the State Highway Department for consideration.

### Financial

At the end of the fiscal year on June 30, 1940, the status of the District revenues and expenditures was on slightly better than a break-even basis. A serious shortage of revenue resulted from the first year's operations, attributable to extreme competitive conditions, inadequate approaches and other factors. □

SIXTY YEARS OF SERVICE

# Highlights from Annual Reports

## YEAR: 1940 — 1941

The fourth fiscal year of Golden Gate Bridge operations, July 1, 1940 to June 30, 1941, established new records in volume of traffic and revenue. A total of 4,764,758 vehicles crossed the Bridge generating revenue of $2,282,213.58.

A major portion of this increased revenue was attributable to augmented interurban bus transportation service inaugurated by Pacific Greyhound Line following termination of passenger-ferry service on March 1, 1941. With an exceptionally fine interurban bus service in operation, offering an attractive schedule, it is to be expected that considerable numbers of casual, as well as commuter users of the Bridge, will find it either more convenient or more economical to use the buses.

### Non-Revenue Government Vehicles

Under this classification is included all vehicular traffic of the Army and Navy, including civilian employees of the Army and Navy traveling on official business, Army and Navy personnel, active or retired, and their dependents, whether on official business or pleasure, and all Army and Navy-owned vehicles or equipment. From and after June 12, 1941, and on a steadily increasing scale, the classification also includes official travel of employees of all other Federal Government agencies.

The amazing total of 382,077 vehicles in the toll-free Government classification were passed in the year 1940-41, an increase of 75.9% over the previous year. The continued rapid increase in volume of Non-Revenue Government vehicles has created growing concern over the cost and inconvenience, both to the District and to toll-paying traffic, of handling the burden. Furthermore, there is no indication of a diminution in the rate of increase.

As a result, the taxpayers of the Golden Gate Bridge and Highway District, compromising the Counties of San Francisco, Marin, Sonoma, Del Norte and portions of Napa and Mendocino, together with all toll-paying users of the Bridge, are bearing more than their just share of the cost of Federal Government activities.

The prospects of obtaining relief from this burden of toll-free traffic, either in the courts or from Congress, appear much less favorable than at this time last year.

### Maintenance

Expenditures for current maintenance during the year amounted to $148,865.28, of which $101,788.20 was for painting bridge and approach structures. Painting continues to be the largest single item of expense in the operation and maintenance of the Bridge. The very large quantity of steel (80,000 tons), and the severe exposure of the Golden Gate to salt-water spray and fog, will make it necessary to continue year-round painting crew to provide constant protection to exposed steel surfaces. All exposed steel must be repainted every four or five years. The first complete repainting should be finished in the fiscal year 1941-42.

---

*In the rest of the World . . . .*

## HISTORICAL SNAPSHOT

**Events.** *Office of Price Administration (OPA) establishes price controls in U.S. "Manhattan Project" (to develop atomic bomb) begins.*

**Sports.** *Jackie Robinson breaks into major league baseball. Lou Gehrig dies.*

**Music.** *Richard Addinsell, Warsaw Concerto; William Schuman, Symphony No. 3.*

**Literature/ Drama.** *Franz Werfel, Song of Bernadette; Noel Coward, Blithe Spirit; Richard Wright, Native Son.*

**Film.** *Citizen Kane, Orson Welles; The Maltese Falcon, dir John Huston; Nous les Gosses.*

**Deaths.** *James Joyce, Sherwood Anderson, Virginia Woolf, Kaiser Wilhelm II.*

**Everyday Life.** *Mt. Rushmore Memorial completed. Number of U.S. private autos: 38.8 million. CBS begins regular TV broadcasts. 'Cheerios' breakfast cereal introduced.*

1941

The District has always been keenly aware of its fiduciary responsibilities. In the 1941 Annual Report it outlined the payment schedule to retire the original construction bonds thusly:

| Period | Maturity Each Year | Maturities | Five-Year Period Interest | Total |
|---|---|---|---|---|
| 1942–1946 | $ 200,000 | $ 1,000,000 | $ 7,509,000 | $ 8,509,000 |
| 1947–1951 | 400,000 | 2,000,000 | 7,205,750 | 9,203,750 |
| 1952–1956 | 800,000 | 4,000,000 | 6,596,500 | 10,596,500 |
| 1957–1961 | 1,200,000 | 6,000,000 | 5,547,000 | 11,547,000 |
| 1962–1966 | 1,600,000 | 8,000,000 | 4,071,500 | 12,071,500 |
| 1967–1971 | 2,800,000 | 14,000,000 | 1,821,000 | 15,821,000 |
| Totals | | $35,000,000 | $32,748,750 | $67,748,750 |

A constant, systematic and careful inspection is made of the Bridge structure, approaches and appurtenances and every necessary measure is take to insure proper maintenance of all properties.

## Highways and Approaches

*[Even four years after opening, the Bridge was still beset by problems of adequate access from the south (San Francisco) and north through San Rafael to the then-popular summer recreation areas in Marin, Sonoma and Napa counties. Also, their continuing effort to create a sea-level alternate road to relieve trucks of the Waldo Grade climb, was being met with gritty denial by the tiny fishing village of Sausalito. In this annual report, the Officers and Directors of the Bridge apparently decided to increase the pressure a bit.]*

Every possible effort was made during the year to promote the construction of an approach road through the Town of Sausalito which would provide a low-level lateral between Waldo Point and the Bridge. The District expended approximately $400,000 on that part of this route extending one and a half miles from the Bridge to the city limits of Sausalito.

There is a four-lane highway from Waldo Point to the northerly limits of Sausalito. But the narrow, crooked streets within the town prevent the use of the route by trucks and buses to avoid the two miles of 6% grade on the Waldo (main) approach. At the end of the fiscal year, efforts were being directed towards obtaining the allocation to this project of Federal funds being appropriated for defense roads. This route is a vital link in highways connecting numerous Army, Air Force and Navy posts in the Bay Area, and a large volume of strictly military traffic would be using it at present if it were completed.

Progress on the project of obtaining and clearing the right-of-way for the widening of Lombard Street in San Francisco has been made by the City and County of San Francisco, although actual work on the roadway has not commenced at the end of the year.

In the meantime, rather serious traffic congestion has developed on Bay Street, east of Van Ness Avenue. While this congestion may be relieved somewhat when Lombard is widened, there still will be a serious need for a direct route between the vicinity of Lombard and Van Ness and the area served by Columbus Avenue. A tunnel through Russian Hill has been mentioned frequently in connection with the problem at this point. Lombard Street is required to accommodate nearly all traffic of the Marina District, the Presidio and the Bridge, and accordingly, a more direct connection with the downtown district is vitally needed.

The elimination of the bottleneck heretofore existing in the City of San Rafael was much nearer to realization at the end of the year. Work on contracts awarded by the State Highway Department for the construction of an elevated freeway was scheduled for completion about September 15, 1941. Improvement work on the highway south of San Rafael likewise was scheduled to continue through the summer. □

# SIXTY YEARS OF SERVICE
# Highlights from Annual Reports
## YEAR: 1941 — 1942

[EDITOR'S NOTE: The fifth annual report of the GOLDEN GATE BRIDGE AND HIGHWAY DISTRICT, presented by General Manager James E. Rickets, contained only 15 short paragraphs. It was typewritten on a few sheets of paper and modestly bound between two orange covers.

In the paragraphs reproduced here, two things are clear. The first is that a new set of priorities are obviously driving the administration and maintenance of the Bridge. Gone is the glorious detail of traffic and revenue. In its place is a no-nonsense report that reflects the dedication of the District to the new, more global business at hand.]

Many important national and international events occurred during the fifth fiscal year of Golden Gate Bridge and Highway District operations, July 1, 1941 to June 30, 1942, which had a definite bearing on the year's volume of traffic and revenue.

General economic conditions resulting from entry of the United States into war have been reflected in truck, passenger car, bus and pedestrian traffic and income.

The United States Government, Office of Censorship, Washington D.C., has requested that the Golden Gate Bridge and Highway District refrain from publishing any statistics concerning the number and types of vehicles crossing the Golden Gate Bridge.

In compliance with this request, the Bridge District has omitted from this annual report all statistical data concerned the traffic.

The Bridge District appealed the decision in the suit brought against it by the United States Government. The adverse court decision resulted in added toll-free use of the bridge by military and Federal personnel and their dependents. This traffic has increased tremendously due to rapidly expanding wartime activities and the fact that the Bay area is fast becoming one of the nation's busiest war industry, and army and navy centers.

While there has already been some noticeable lessening of traffic due to rubber shortage and reduction of non-essential traffic, inauguration of a nation-wide gasoline rationing program will unquestionably result in further retarded transient customer revenues.

In anticipation of such possible revenue reductions, plus possible future restrictions which may be imposed by the campaign has been launched in cooperation with the Redwood Empire Association to enact legislation whereby the Federal Government will financial assist in equalizing wartime revenue losses thereby relieving counties included in the Bridge District of such responsibilities.

Due to resignations, there were changes in the offices of the general manager and auditor during the fiscal year and numerous personnel changes have been made in connection with men entering the military service or employment in defense industries without interruption to the operations of the District or service to the public.

## In the rest of the World . . . .
# HISTORICAL SNAPSHOT

**Events.** *Civilian auto production ceases in U.S., coffee, sugar, rationed, also gasoline limited to 3 gals/wk. Henry J. Kaiser becomes major shipbuilder, 4 West Coast Plants (including Sausalito) build Liberty ships. Frank Lloyd Wright designs Guggenheim Museum, NY.*

**Drama.** *Jean Anouilh, Antigone, Thornton Wilder, The Skin of our Teeth.*

**Music.** *Frank Sinatra's first stage appearance in New York. Bing Crosby introduces "White Christmas." Glen Miller enters Air Force.*

**Films.** *Casablanca; The Magnificent Ambersons, dir, Orson Welles; Woman of the Year, starring Tracy and Hepburn.*

**Sports.** *39th World Series won by Cards over Yankees, 4-1. Top golf money winner, Ben Hogan with $13,143.*

**Everyday Living.** *Entire U.S. goes on "War Time" — as in daylight savings time, all year long.*

# 1942

**San Francisco Growth.** *Along with the rest of the United States, San Francisco was coming out of the Great Depression during and after the building of the Golden Gate Bridge. In 1942 wartime restrictions slowed the growth of housing. It would be several years before the rest of the open land around Mt. Davidson was filled with San Francisco's infamous "ticky-tacky little boxes." The cross (center) is the site of Easter sunrise services and was owned by the city until 1997 when it was sold in response to a court order that determined it was a violation of the U.S. Constitution which mandates a separation of church and state.*

## Maintenance

The sum of $150,366.23 was expended for the twelve month period to adequately maintain the Golden Gate Bridge proper and approach structures. Of this amount, $95,057.74 was expended for painting operations, which included the Marin Tower and the main suspended span.

Scientific instruments for determining the direction and velocity of the wind and recording vibration have been purchased and mounted at the midspan of the 4200-foot suspended span. A twenty-four hour continuous record is maintained from the two wind instruments and the seismograph. From these records, useful and accurate information is obtained as to the various movements caused under different wind conditions. Up to this time no detrimental results have been recorded.

All governmental requirements have been met relative to the maintenance and operation of the bridge since the declaration of war.

A military guard is being furnished for the protection of the bridge, approaches and cable anchorages. ☐

## SIXTY YEARS OF SERVICE

# Highlights from Annual Reports

## YEAR: 1942 — 1943

*[Editor's Note: This year the District Annual Report was equally brief — two and one-half typewritten pages (plus the financial report), reproduced on low-quality paper, bound by simple orange covers. Problems were stated simply and honestly and, careful reading will reveal, with the pride of a job well done under sometimes trying circumstances.]*

In its first full wartime year, and in its sixth fiscal year, July 1, 1942 to June 30, 1943, of operation, the Golden Gate Bridge established new records in revenue, which directly reflected the heavy traffic of the industrial war effort and the efficient administration of the Board of Directors of the Golden Gate Bridge and Highway District.

Despite war conditions and restrictions on traffic due to gasoline, tire and automobile rationing, the Bridge District ended the year in the strongest financial position since its opening on May 27, 1937.

The Board continued its policy of keeping the operation and maintenance costs as low as possible, consistent with proper safeguarding of the District's property and efficient operation of the Bridge. As a result, the District has been able to set aside out of surplus $400,000 to meet bonds maturing in 1944 and 1945.

The most noteworthy factor affecting operation of the Bridge during the fiscal year 1942-1943 was the opening of Marinship, a great wartime shipyard at Sausalito, on the north shore of San Francisco Bay. Most of the 17,500 employees of Marinship reside in San Francisco and use the Golden Gate Bridge going to and from work.

A serious problem keeping the Bridge operating staff at full strength and efficiency was imposed during the year on account of the number of employees who joined the armed forces, or became engaged in essential war industries.

### Maintenance

The cost of maintaining the Golden Gate Bridge structure, approaches and pavements, for the twelve-month period, amounted to $126,931.35. Of this sum, 64.26 per cent, or $81,563.90 was expended for carrying out the Bridge painting program, which included painting of the arch span and the San Francisco truss span.

For six years, painting of the Golden Gate Bridge has been carried on continuously by a staff of painters whose experience qualifies them especially for bridge painting. Many of the employees in the painting force were on the project from the beginning, and they are trained to work on the various types of design on the Golden Gate Bridge.

The various paints purchased are prepared in accordance with the paint formulas established for use on the Bridge to protect the steel surfaces against the very severe weather conditions peculiar to this locality.

Curtailment of normal Bridge maintenance schedules due to the war effort, which has affected the supply of both labor and material, was to be expected, but in spite of severe regulations, the Bridge property is being maintained at a satisfactory standard.

---

*In the rest of the World . . . .*

## HISTORICAL SNAPSHOT

**Events.** *A polio epidemic strikes the U.S., killing 1151, crippling thousands. A minimum 48-hour week declared for war plants. Rationing of shoes (3 pairs/yr), canned goods, meat, fat and cheese begins.*

**Music.** *Oklahoma and Carmen Jones open on Broadway.*

**Drama.** *Bertolt Brecht, The Life of Galileo; Jean-Paul Sarte, Les mouches.*

**Literature.** *Antoine de Saint-Exupery, The Little Prince.*

**Films.** *For Whom the Bell Tolls, starring Gary Cooper, Ingrid Bergman; Lassie Come Home, starring the male dog, Pal.*

**Radio.** *Edward Noble founds the American Broadcasting Company, ABC.*

**Everyday Life.** *Zoot suits become high fashion in the U.S.*

# 1943

## FOR THE RECORD

*From Opening Day 1937, the Golden Gate Bridge's main span of 4,200 feet (1,280 m) was the longest in the world - until the Veranzano Narrow Bridge was* *constructed twenty-seven years later. Now even that record has been eclipsed. Here are the major suspension bridges in the world and their ranking.*

| Ranking | Bridge Name | Length of Ctr Span (m) | Remarks |
|---|---|---|---|
| 1. | Humber Bridge | 1,410 | UK, completed 1981 |
| 2. | Verrazano Narrows Bridge | 1,298 | USA, completed 1964 |
| 3. | *Golden Gate Bridge* | *1,280* | *USA, completed 1937* |
| 4. | Mackinac-Straits Bridge | 1,158 | USA, completed 1957 |
| 5. | South Bisan-Seto Bridge | 1,100 | Japan, completed 1988 |
| 6. | Bosphorus Bridge | 1,074 | Turkey, completed 1973 |
| 7. | George Washington Bridge | 1,067 | USA, completed 1931 |
| 8. | Ponte 25, April Bridge | 1,013 | Portugal, completed 1966 |
| 9. | Forth Road Bridge | 1,006 | UK, completed 1964 |
| 10. | North Bisan-Seto Bridge | 990 | Japan, completed 1988 |
| 11. | Severn Bridge | 988 | UK, completed 1964 |
| 12. | Shimotsui-Seto Bridge | 940 | Japan, completed 1988 |
| 13. | Ohnaruto Bridge | 876 | Japan, completed 1985 |
| 14. | New Tacoma Narrows Bridge | 853 | USA, completed 1950 |
| 15. | Innoshima Bridge | 770 | Japan, completed 1983 |
| 16. | Angostura Bridge | 712 | Venezuela, competed 1967 |
| 16. (tie) | Kammon Bridge | 712 | Japan, completed 1973 |

The installation of eight additional scientific vibration recording instruments, to be located at different points on the main suspended span and the two side spans, will provide an actual daily record for the entire length of the structure. Daily vibration-chart records for the fiscal year reflect nothing more than normal movement of the Bridge.

### Financial

Notwithstanding war conditions, earnings of the Golden Gate Bridge and Highway District for the fiscal year ended June 30, 1943 were very satisfactory. Gross revenues amounted to $2,579,981.91. □

SIXTY YEARS OF SERVICE

# Highlights from Annual Reports

YEAR: 1943 — 1944

Completing its seventh fiscal year of operation for the period July 1, 1943 to June 30, 1944, the Golden Gate Bridge maintained its strong financial position in the face of adverse war-time conditions affecting automobile traffic. Continuation of gasoline, automobile and tire rationing, as well as a decrease in the volume of motor vehicles now in use, had their effect on Bridge revenues.

Outstanding in the year's developments affecting the Bridge traffic and revenue was the passage by the Congress of the United States of legislation which curtailed toll-free Government traffic crossing the Bridge.

The purpose of this legislation was to afford the Bridge District a measure of relief from the financial burden imposed upon it as a result of the original War Department permit for the construction of the Bridge, which required that all Government traffic over it would be toll free. A War Department permit to build the Bridge was necessary because both ends of the Bridge and its approaches cross military reservations.

The legislation, as finally passed and signed by President Roosevelt, limited toll-free Government traffic to Army and Navy personnel and civilian employees of the Army and Navy while traveling on official business. It eliminated free tolls for all other Federal agencies, and for dependents of Army and Navy personnel.

Since this law took effect, the toll-free Government traffic has dropped from some 2,200 vehicles per day to less than 800 per day.

## Personnel

Despite manpower problems, the Bridge operation staff was maintained at virtually full strength during the fiscal year. Some employees became affiliated with the armed services or resigned to engage in essential war industries but a great majority remained with the District regardless of advantages offered them in private industry. Credit is due these for their interest in their work and the excellent results obtained in operating a maintaining the huge structure.

After considering various pension plans submitted by private companies, The Board of Directors, on account of the substantial savings in cost to the District and to the employees, decided to sponsor legislation permitting employees of the District to become members of the California State Retirement System. The necessary legislation having been enacted at the 1943 session, the employees by a large majority voted to join the System and the Board of Directors entered into a contract whereby employees of the District became members of the System March 1, 1944, with full credit for prior service.

---

*In the rest of the World . . . .*

## HISTORICAL SNAPSHOT

**Events.** *Roosevelt elected for 4th term as U.S. president. D-Day June 6. In France, provisional goverment allows women to sit in parliament. First non-stop flight London-Canada. Kidney machine invented.*

**Music.** *Prokofiev, Symphony No. 5; Richard Strauss, Die Liebe der Danae; Musical, On The Town, Comden & Green, L. Bernstein; Capitol Records formed.*

**Drama.** *Tennessee William, The Glass Menagerie.*

**Books.** *Joyce Cary, The Horse's Mouth; Somerset Maugham, The Razor's Edge.*

**Films.** *Arsenic and Old Lace, dir Frank Capra; Lifeboat, dir Alfred Hitchcock.*

**Sports.** *In the 41st World Series, the St. Louis Cardinals defeated the St. Louis Browns 4 games to 2.*

**Everyday Life.** *Meat rationing ends, except for steak.*

# 1944

A look at the investments made by the Golden Gate Bridge and Highway District during the fiscal year reveals a red, white, and blue patriotic portfolio.

### INVESTMENTS

June 30, 1944

| | Interest Rate | Maturity | Cost | Maturity Value |
|---|---|---|---|---|
| U.S. Savings Bonds Series F | If held to maturity - 2.53% | 1953-1956 | $ 299,700 | $ 405,000 |
| U.S. Treasury Bonds | 2% | Sept 15, 50/52 | 250,000 | 250,000 |
| U.S. Treasury Bonds | 2 1/4% | Sept 15, 56/59 | 200,000 | 200,000 |
| U.S. Treasury Bonds | 2 1/2% | Sept 15, 64/69 | 250,000 | 250,000 |
| U.S. Treasury Bonds | 2 1/2% | Sept 15, 64/69 | 100,000 | 100,000 |
| U.S. Treasury Bonds | 2 1/2% | Sept 15, 65/70 | 300,000 | 300,000 |
| | | | $1,399,700 | $1,505,000 |

## Sausalito Lateral Extension Project

During the fiscal year the Bridge management, under instructions from the Board of Directors, held several conferences with members of the State Highway Commission, public officials and citizens, interested in construction of the Sausalito Lateral extension project.

There was considerable progress in promoting the project, which has been given the endorsement of numerous public bodies and civic organizations in the District.

## Vibration and Wind Recorders

Daily vibration-chart readings, as recorded on vibration and wind-recording instruments located on the Bridge, reflected no more than normal movements of the Bridge during the year.

The United States Bureau of Public Roads became interested in this apparatus and an agreement between the bureau and the Bridge District was signed, continuing for another year the agreement which has been in effect since May 15, 1943, whereby both parties share equally the cost of constructing additional vibration recorders for installation on the Bridge.

## Financial

Earnings of the Golden Gate Bridge and Highway District for the fiscal year ended June 30, 1944 were very satisfactory. Gross revenues amounted to $2,495,220.32 compared with $2,579,981.91 for the previous fiscal year.

## Maintenance

The maintenance and repair cost on the Golden Gate Bridge for the seventh year of operation amounted to $128,563.03, which is $20,008.97 less than the amount appropriated for such purposes. The anticipated expenditures were restricted due to shortage of manpower and to government priorities controlling and allocating material to preferred maintenance only. The Sausalito Lateral pavement also reflects continuous use by trucks, buses, and Government vehicles enroute to Marinship, Mare Island, Hamilton Field, and other war plants and bases. The Federal Government will be asked to allocate funds, set up for such contingencies, to repair the 1-1/8 miles of road owned by the District.

It is estimated that a large sum of money will be required to rehabilitate the approach road pavements during the post-war era. This condition is typical all over the west due to the movement of heavy motorized military equipment and trucks. After the war, when restrictions are removed and normal conditions prevail, a larger scale program of repairs and maintenance will be considered. □

SIXTY YEARS OF SERVICE

# Highlights from Annual Reports

YEAR: 1944 — 1945

[This year marks the first annual report by the District to include photographs of the Board and Officers. It is also the first time color was added to the document, in this case a limited use of red borders and lines for decoration. This Report also marked a return to a printed (not typewritten) format. The paper was glossy and was enclosed in a two-color cover.]

The GOLDEN GATE BRIDGE achieved new records of traffic and revenue during 1944-45, the eighth fiscal year of operation, ended June 30, 1945. As a result, the Bridge District continued to strengthen its financial position to a marked degree. The Bridge District not only met all obligations, as it has from its inception, but in addition during the year, earned $554,982.65.

While wartime restrictions affecting motor vehicles and travel generally remain in effect, nevertheless bridge traffic and revenue were moving progressively forward as the fiscal year ended.

The end of the war in Europe in May, 1945, brought an immediate upswing in the number of vehicles crossing the bridge, caused primarily by the modification of government rationing limitations on gasoline and tires.

## Legislation

During the 1945 session of the California state legislature, (Bay Area) assemblymen sponsored a bill authorizing the state to loan $5,000,000 to the Bridge District for construction of a new low-level Marin County approach to the bridge from the north.

Briefly, the measure provide that the State Division of Highways shall construct a modern six-lane highway approach to the bridge from Waldo Point in Marin County after the route and plans and specifications have first been approved by the Bridge District. The approach, when completed, will become part of the primary state highway system.

Another bill of importance to the District directs the State Division of Highways to take over and maintain the Funston Avenue approach, a present a state highway leading to and from the western residential section of San Francisco. The cost of such maintenance, now borne by the Bridge District, is estimated at approximately $15,000 annually.

## Retirement

Mr. W. W. Felt, Jr., secretary of the District and oldest employee in point of service, retired voluntarily because of ill health on April 30, 1945. He had been employed as secretary since the formation of the District on March 13, 1929.

## Maintenance

Expenditures for the maintenance of the Golden Gate Bridge for the fiscal years 1944-45 were, as in previous years, largely curtailed on account of government restrictions on labor and material diverted to the war effort. To win the war being the nation's first objective, only such maintenance as could not be postponed without risk to the bridge was undertaken.

---

*In the rest of the World . . . .*

## HISTORICAL SNAPSHOT

**Events.** *Word War II end in Europe May 8, in Pacific, August 15. A B-25 bomber flew into the 78th floor of the Empire State Building in New York.*

**Music.** *Bartok, Piano Concerto No. 3; Carousel, Rodgers & Hammerstein; Hit song, 'Till the End of Time, Perry Como.*

**Books.** *John Hershey, A Bell for Adano; George Orwell, Animal Farm; John Steinbeck, Cannery Row; Richard Wright, Black Boy.*

**Films.** *The Lost Weekend, dir Billy Wilder; Open City, dir Roberto Rossellini.*

**Sports.** *Rocky Graziano scored five knockouts at Madison Square Garden. Golf's top money winner was Byron Nelson, $63,335.66. NFL: Cleveland Rams defeated Washington Redskins, 15-14.*

**Everyday Life.** *Rationing ended for all meat, butter, shoes, tires. Biro Ballpoint pens intoduced in U.S. "Bebop" music sweeps the country.*

# 1945

**The Quiet Before the Storm.** *As America returned to a peacetime economy, the Golden Gate Bridge enjoys a lonely moment, as if taking a deep breath in preparation for what lies ahead. At age eight years, it is about to enter a period of unprecedented growth. This is the last quiet moment the span will ever know.*

The effect of the war on labor and materials has greatly complicated the problems entailed in maintaining the structure. When peace is won and conditions are again stabilized, paint products and methods of application will be constantly studied to secure the best results.

It is anticipated that many items in the post-war budgets will be necessarily increased in order to provide for such maintenance as could not be performed during the war period.

### Financial

Revenues of the Golden Gate Bridge and Highway District for the fiscal year ended June 30, 1945, showed a marked increase over previous years. Total gross revenues for the year amounted to $2,710,710.75, an increase of $215,490.43 over the year 1943-44. □

SIXTY YEARS OF SERVICE

# Highlights from Annual Reports

## YEAR: 1945 — 1946

THE FISCAL YEAR 1945-46 was the most successful year of operation of the Golden Gate Bridge and Highway District; both traffic and revenue reached an all-time high.

The traffic trend upwards, which began in May 1945 immediately after the termination of hostilities in Europe, continued progressively as the government removed all rationing restrictions on gasoline, and on tires and new automobiles following the surrender by Japan.

It is gratifying to report that the Golden Gate Bridge contributed significantly to the magnificent war achievements of the San Francisco Bay area, one of the largest war production centers in the world, and San Francisco, the greatest port of embarkation in the country.

A daily average of 18,198 vehicles passed over the bridge during the year. The average daily revenue was $9,807.

### New Marin County Approach

In recent months splendid progress has been made on an intensive campaign to construct a new Marin County approach to the bridge and to realign and otherwise improve feeder roads, which will aid the development of nearby communities and stimulate the flow of traffic.

The building of a new Marin County approach (is the result of a law which became effective September 15, 1945.) Under this legislation, the State of California was authorized to lend the Bridge District the sum of $5,000,000 for the construction of a new six-lane divided highway from the north end of the Golden Gate Bridge to a point at or near Waldo Point on U.S. Highway 101 in Marin County.

*[On April 26, 1946 the State Division of Highways submitted plans and estimates to the District for approval. The cost estimates exceeded the five-million dollar appropriation and were referred to a special Bridge committee, who later requested the State to make further surveys.]*

### Financial

Total revenues of the Bridge District for the fiscal year ended June 30, 1946 amounted to $3,579,521.18, an increase of 32% over the previous year. Revenues from transient passenger automobiles showed an increase of 56.7%. Commercial and residential development in Marin County has contributed in large measure to an increase of 12% in truck revenue and 26.7% in commute sales over the previous fiscal year, notwithstanding loss of revenue due to the termination of shipbuilding operations at Marinship in November of 1945.

---

*In the rest of the World . . . .*

## HISTORICAL SNAPSHOT

**Events.** *John D. Rockefeller, Jr. gives U.N. $8,500,000 to purchase property for permanent UN headquarters. U.S. conducts atomic bomb tests at Bikini Atoll in Pacific. ENIAC, "world's first computer" set up in 30-by-60 (feet) room, weighs 30 tons, in Philadelphia.*

**Books.** *Robert Penn Warren, All the King's Men.*

**Drama.** *Eugene O'Neil, The Iceman Cometh; Jean-Paul Sartre, Morts sans sepulture.*

**Sports.** *Golf's top money winner: Ben Hogan, $42,566.36.*

**Films.** *The Best Years of Our Lives, dir William Wyler; The Killers, Burt Lancaster's debut; Shoeshine, dir Vittorio de Sica, Les Portes de la Nuit, dir Marcel Carne.*

**Everyday Life.** *The first bikinis create a sensation when previewed in Paris. TIDE becomes first consumer product to replace soap. First drive-in bank opens.*

# 1946

**WHAT ARE THESE PEOPLE DOING?** *It's obvious where they are, even if you don't know what they're* *doing. Hint: flash forward forty years. If you can't wait for the answer,* *turn to page 265.*

### Maintenance

Normal maintenance of the bridge structure and facilities, which had been partially curtailed during the war years because of the shortage of labor and supplies was restored. Despite maintenance problems, all bridge property was constantly inspected and protected during the war, and not a single break-down of facilities occurred. Engineers have pronounced the bridge structure in first class condition.

To insure the proper protection of exposed steel surfaces of the bridge, the painting force was enlarged in the last year to a complement of 30 experienced men. A total of 30,000 tons of steel was painted. Experiments have been conducted, and are continuing, with samples of new protective coating materials for the purpose of testing their rust-resisting qualities in comparison with paint now being used.

### Personnel

At the end of the year there were 102 employees in the Bridge District service. □

SIXTY YEARS OF SERVICE

# Highlights from Annual Reports

YEAR: 1946 — 1947

## The Tenth Anniversary

The Golden Gate Bridge, the world's longest single span suspension bridge, has been in operation ten years on May 27, 1947. In the ten-year period, the bridge, which directly links San Francisco and the northern counties, spanning the world-famed Golden Gate, the entrance to San Francisco harbor, has proved its value in war and peace both as a public convenience and vital necessity.

A total of 7,458,424 vehicles crossed the bridge during the year 1946-47, such numbers being more than double the 3,311,512 vehicles which crossed the bridge during the first full fiscal year of 1937-38.

The daily average of vehicles using the span last year was 20,434 with a daily average revenue of $10,589.

It is predicted that by 1951 annual traffic volume on the bridge will have reached a total of over 9,000,000 vehicles.

On the recommendation of the Finance Committee, the Board authorized the elimination of tolls for extra passengers in all vehicles, except commute automobiles, common carriers and toll-free government vehicles. It also reduced the toll for light delivery trucks bearing passenger car license plates from 60-70 cents to 50 cents.

## Financial

Total revenues of the Bridge District for the fiscal year ended June 30, 1947 amounted to $3,864,817, an increase of 8% over the previous year.

## Maintenance

Maintenance of the Golden Gate Bridge was curtailed to some extent during the war years, due to the scarcity of materials. Since the war, all that is possible to accomplish has been done to ensure the protection of the structure against the elements.

Regular inspections of the property of the District are made by competent personnel. To safeguard the one hundred and seven thousand tons of structural steel and cable wire against corrosion, thirty full time experienced bridge painters have been employed. This force applies over 10,000 gallons of paint to approximately thirty-two thousand tons of steel each year.

Various manufacturers' sample of protective coatings for steel surfaces have been given field exposure tests at the bridge site to determine their rust-resistant qualities as compared with the paint now being used on the structure. Qualified engineers have expressed the opinion that none is better than the paint now used and declare the structure to be in a remarkable state of preservation.

*[Following the early struggle to establish the District and to span the gate with "the bridge that couldn't be built," and early problems with rate-cutting ferries, poor road approaches, and a high incidence of toll-free Government traffic, the Bridge celebrated its 10th birthday with a mixture of optimism, caution and honestly-earned pride.*

*The Report, for the first time, looks ahead to the constantly-growing traffic and sounds a distant early warning as to the possibility that the Bridge might someday approach vehicular capacity and face the challenges that would bring.]*

---

*In the rest of the World . . . .*

## HISTORICAL SNAPSHOT

**Events.** *Fire at docks in Texas City, TX kills 500. New constitution in Japan allows women to vote. The U.S. Airforce is established. de Gaulle organizes the RPF. Odum flies around the world in 73 hours, 5 minutes. Pan Am makes first globe-circling commercial flight, fare $1,700.*

**Music.** *Brigadoon and Finian's Rainbow.*

**Drama.** *Arthur Miller, All My Sons; Tennessee Williams, A Streetcar Named Desire; The Actors' Studio founded by Elia Kazan. Tony Awards established.*

**Born.** *Salman Rushdie, David Mamet.*

**Deaths.** *Henry Ford, Willa Cather, Ettore Bugatti.*

**Sports.** *Jackie Robinson signs with Brooklyn Dodgers.*

**Broadcasting.** *"Voice of America" founded. Pres. Harry Truman makes first presidential TV address.*

**Everyday Life.** *Microwave ovens introduced. First reported citings of UFO in U.S., over Kansas.*

**The Golden Gate Bridge has been in operation ten years.**

# 1947

# THE GOLDEN GATE BRIDGE

By Joseph B. Strauss

*I am the thing that men denied,*
*The right to be, the urge to live;*
*And I am that which men defied,*
*Yet I ask naught for what I give.*

*My arms are flung across the deep,*
*Into the clouds my towers soar,*
*And where the waters never sleep,*
*I guard the California shore.*

*Above the fogs of scorn and doubt,*
*Triumphant gleams my web of steel;*
*Still Shall I ride the wild storms out,*
*And still the thrill of conquest feel.*

*The passing world may never know*
*The epic of my grim travail;*
*It matters not, nor friend or foe —*
*My place to serve and none to fail.*

*My being cradled in despair,*
*Now grow so wondrous fair and strong,*
*And glorified beyond compare*
*Rebukes the error and the wrong.*

*Vast shafts of steel, wave-battered pier,*
*And all the splendor meant to be;*
*Wind-swept and free, these, year on year,*
*Shall chant my hymn of Victory!*

Another of Strauss' poems. Written ten years ago, in 1937, this time the poet-engineer seems to take a last look back at the problems and situations he endured to bring the span from dream to reality, and says in heroic verse, "The struggle is over; the Bridge speaks for itself."

---

Ten vibration instruments which were installed at various points on the bridge in 1945-46 have recorded valuable information relating to the movement of the bridge. These data are used in connection with the investigation and research work being carried on in cooperation with the United States Public Roads Administration for the purpose of determining the design and limitations which must be placed on suspension bridges. Never has the movement approached a degree such as would cause an interruption of the flow of traffic over the bridge.

## Legislation and Public Relations

Since 1943 the Bridge District has maintained a legislative representative during the regular and special session of the State Legislature at Sacramento, and the results have justified the action of the board in adopting this policy.

Close cooperation exists between the Bridge District and the State Legislature and all agencies of the City, County, State and Federal governments.

Likewise, a policy of constructive public relations has been maintained by the bridge board and management which recognizes that its first responsibility is to the public. Therefore, the District has endeavored to provide information which will contribute to the fullest possible understanding of the board's position and operation policies for the public convenience.

## New Marin County Approach

During the last fiscal year, continued progress was made on the new Marin County approach project, designed to provide a more adequate approach to the Golden Gate bridge from the north to accommodate traffic needs now and in the future.

Because of the vital importance of the project to the bridge and the entire area lying to the north, the bridge board, through special committees, has give most serious study to all proposals. The studies are being continued and it is hoped that some concrete action on a selected route may be taken during the coming fiscal year, in cooperation with state highway officials. ☐

## SIXTY YEARS OF SERVICE

# Highlights from Annual Reports

## YEAR: 1948 — 1949

### Report of the President

The fiscal year 1948-49 was the most successful in the history of the Golden Gate Bridge since it was opened to traffic in May 1937.

To report that is a real pleasure. It is tangible evidence of the strength of the Golden Gate Bridge and Highway District. It reflects sound planning — intelligently and earnestly administered to insure financial stability in operations.

The financial statement, as printed in this twelfth annual report [but not included here — Ed.], clearly shows the unusually strong and sound position of the District.

Proudly we report the 1948-49 net income of $2,053,830, (compared with $1,962,596 for the prior year).

### Engineering Survey

During the last year, work has been diligently pursued in a comprehensive engineering survey of the bridge structure, approaches, cables and piers and other facilities.

This has included exhaustive tests and borings on the sites of the north (Marin) and south (San Francisco) piers, which support the giant bridge towers. The United States Coast and Geodetic Survey has cooperated by conducting a special survey in the vicinity of the Golden Gate Bridge for the purpose of determining the bottom contour and various depths and curves near the bridge piers.

The south pier is protected by a concrete fender, built during the original bridge construction. The north pier, which is erected on solid rock, has no such protection as the engineering board determined at the time of building that its position on the shore line made it unnecessary. Up until now, the north pier has been protected by tons of rip rap.

[Rip rap is a wall made of broken stones — in this case, very large ones — thrown together irregularly or loosely. — Ed.]

While some of the rip rap has been washed away by the buffeting of waves from the open sea, engineers have already found by inspection and preliminary tests that the structural qualities of the pier and its foundation have not been impaired in any way.

### Report of the General Manager

During the year, a total of 8,347,790 vehicles crossed the bridge as compared with 7,923,700 in the previous year or an increase of 5.35 per cent.

### Traffic Analysis

There has been a continuing upward trend in development of residential sections of Marin County and its is reasonable to expect that with such development the increase in traffic over the Bridge will be greater in the commutation bracket than in the cash toll bracket.

[This year's highlights reflect a change in reporting style. For the first time the President of the Board makes a report, dealing primarily with policy, legal, and financial matters (much of which we have chosen not to use). This was followed by a General Manager's Report, whose details are richer and whose concerns are a continuation of facts we've been following throughout these Highlights.]

---

In the rest of the World . . . .

## HISTORICAL SNAPSHOT

**Events.** *North Atlantic Treaty Organization (NATO) formed. UN headquarters dedicated in New York. Minimum wage raised to 75 cents/hr. Cortisone discovered.*

**Music.** *Gentlemen Prefer Blondes and South Pacific open on Broadway. RCA introduces first 45rpm records.*

**Books.** *Paul Eluard, Une Lecon de morale (poems); George Orwell, Nineteen Eighty-four.*

**Sports.** *Joe Louis retires after 25 title defenses. Top money-winner in women's golf is Babe Didrikson Zaharias, who wins $4,650. New York defeated Brooklyn 4 games to 1 in 46th World Series. In NCAA basketball championships, Kentucky defeated Oklahoma State, 46-36.*

**Everyday Life.** *"Silly Putty" was invented by accident.*

# 1949

For the two previous years, when a charge was made for each individual bus passenger, the records indicate a decided decrease in the volume of passengers carried by mass transportation vehicles. This no doubt was due in considerable measure to the increasing availability of automobiles for personal travel to and from work and would account, to some extent, for a part of the increase in automobile use.

Analysis of weekly passenger automobile traffic in both directions shows Sundays to be the highest with 22.7 per cent of the weekly total; Saturday next with 15.7 percent; Friday 13.6 per cent; Monday 12.4 percent; and about 11.7 per cent for each of the remaining days, Tuesday, Wednesday and Thursday.

The average number of persons occupying automobiles ranges from 2.1 on Monday to 2.5 on Saturday and 3.0 on Sunday. On a week day during daylight hours about 40 per cent of the automobiles carry only one passenger each and 33 per cent carry two passengers each.

## Maintenance

The most costly item of maintenance for the world's longest single span suspension bridge is painting. During the last year, a total of over 28,000 tons of structural steel and wire cables was painted, which consisted of the San Francisco approach spans, arch span, main cables, and suspender ropes and assemblies, including eye bars.

Measures have been take to prevent further wear of the concrete roadway pavement expansion joints, especially on the suspended and approach spans, by sealing the joints with a rubber sealing compound.

## Engineering

Studies were initiated into the matter of providing a permanent protective fender around the North (Marin) Pier of the bridge similar to that built during bridge construction around the South Pier. The base of this (north) pier comprises solid reinforced concrete 80 feet wide by 160 feet long, residing on solid rock formations.

## Personnel

Loyal cooperation of the 110 employees of the bridge district has been given the management at all times and was a major contributing factor in the successful and efficient operation of the bridge.

Excellent relations exist with the employees, all of whom work on a five day-forty hour week, which prevails among most public and private employers in the state of California. □

SIXTY YEARS OF SERVICE

# Highlights from Annual Reports

YEAR: 1949 — 1950

## Foreword

THE GOLDEN GATE BRIDGE is playing a dominant part in the phenomenal growth of the state of California.

As a link between San Francisco and the vast Redwood Empire to the north, it has stimulated the development of this productive and scenic northern coastal area, and the growth is significantly reflected in the 1950 federal census for the counties of the Golden Gate Bridge and Highway District.

The Golden Gate Bridge is operated as a public trust by a board of 14 directors of the Bridge and Highway District, which is a political subdivision of the state of California.

The directors are chosen by the Boards of Supervisors of their respective counties and serve for four-year terms each. By state law, they are charged with full responsibility for governing the district, and adopting rules of policy.

On June 30, 1950, the Golden Gate Bridge closed it thirteenth full year of operation with another record year of success.

The outlook for the future is most favorable.

## Report of the President

The highlight of the year's activities was a decision of the board of directors on May 26, 1950 to authorize a reduction in tolls on the Golden Gate Bridge, effective July 1, 1950.

The board's action provided for a reduction in the transient one-trip passenger automobile toll from 50 cents to 40 cents, and a toll charge for trucks by axles, instead of by weight.

In revising tolls, the Finance Committee and the board took into consideration the unusually low toll enjoyed for many years by the regular bridge commuters.

The basic 20-cent toll for regular automobile commuters on the Golden Gate Bridge has been in effect for 12 years and represents a 60 per cent reduction over the transient one-way rate for passenger cars.

The transient rider, who pays a 50-cent cash toll at present —and has since the bridge was opened in 1937 — contributes approximately 62 per cent of the bridge income.

The decision of the board to apply an axle toll rate to trucks, instead of a weighing toll, was based on several factors. The rates adopted were the most equitable that could be applied.

By eliminating the use of scales for trucks, the bridge district will avoid the necessity of costly physical alterations in the toll lanes; and do away with the maintenance, repair and ultimate replacement of scales and electrical auxiliaries at a heavy cost.

In addition, the removal of scales will materially expedite the flow of truck traffic through present truck lanes, and make it possible to use other lanes for trucks when additional lane capacity is required.

The truck operators, including our farmers and business people, bring dairy products, fish, meat, fruit, poultry, lumber, wool and many other products to San Francisco markets over the Golden Gate Bridge while a steady flow of foodstuffs and materials and supplies moves from San Francisco across the bridge to Marin county and other north bay counties.

*[Something new this year — a Forward added to page one of the Report. More than anything else, it represents the Board's recognition of the Bridge's importance to the economic well-being and booming development of Northern California. It says as much without boasting and reminds everyone they execute their duties and public trust with the utmost regard and consideration.]*

---

*In the rest of the World . . . .*

## HISTORICAL SNAPSHOT

The Bridge was conceived and built during an important, expanding time in history. It's birth was difficult; it exists in a real world. *These are a few of the events, near and far, that happened concurrently with its development and growth. As you'll see, from time to time events occuring far from San Francisco Bay held influence over the business of the Golden Gate Bridge.*

**Events.** *The Korean War began when North Korean troops crossed the 38th parallel. U.S. sends ground forces to Korea; President Truman declares National State of Emergency. Rationing ends in West Germany, except for sugar. Illiteracy in the U.S. reaches all-time low of 3.2% of population. Longest U.S. vehicular tunnel — the Brooklyn-Battery Tunnel in New York — opened to traffic.* [**The BART tunnel is longer.**]

**Books.** *Ray Bradbury, The Martian Chronicles; Hemingway, Across the River and Into the Trees.*

**Drama.** *Jean Anouilh, La Repetition ou l'amour puni; William Inge, Come Back Little Sheba; John Van Druten, Bell, Book and Candle.*

**Music.** *Maria Callas debuts at La Scala; Call Me Madam and Guys and Dolls open on Broadway.*

**Deaths.** *George Orwell, Sir Harry Lauder, Edgar Rice Burroughs, Kurt Weill, Vaslav Nijinsky, Al Jolsen, George Bernard Shaw.*

**Films.** *All About Eve, dir Joseph Mankiewicz; Les Enfants terribles, dir Jean Cocteau; Rashomon, dir Akira Kurosawa; Sunset Boulevard, dir Billy Wilder.*

**Everyday Life.** *Charlie Brown makes first appearance in funnies. Smokey the Bear becomes emblem of the National Park Service.*

**The Golden Gate Bridge** — *vehicular traffic was approaching the eight-digit mark.*

## 1950

Lower truck tolls mean lower living costs. The Finance Committee of the bridge board has done a most excellent job in its comprehensive study of the toll situation and deserves the commendation of the public.

### Personnel

Although the bridge district is a government agency, its management efficiency compares favorably with that of the best operated private corporations. The loyal cooperation of the employees has been a major factor in the successful operation of the bridge and the board expresses its appreciation to them.

### Report of the General Manager

The annual bridge revenues passed the $4,500,000 mark for the first time ($4,519,515) in the last fiscal year, and traffic soared toward the 10,000,000 annual figure (9,313,899).

### Toll Reductions

In view of the limited capacity for trucks at the Golden Gate Bridge toll plaza, where there is only one lane available in each direction for weighing trucks, special consideration was given to determine methods by which truck traffic volume could be increased without added congestion.

Under the new axle toll method approved for truck tolls, the necessity of using scales for weighing is removed and thus it will be possible to use more than one lane for trucks if the need arises.

Consideration and study were also given to other means of increasing toll lane capacity, which will be put into effect during the coming fiscal year.

### Traffic

During the fiscal year 1949-50 there was an increase of 9.5 per cent in total vehicles over the previous year. Passenger automobiles, including commutation vehicles, which represent 88.9 per cent of total vehicles increased by 10.2 per cent.

As a means of increasing bridge capacity to meet the upward trend in traffic, physical alterations were made during the year in certain facilities.

The Toll Sergeant's booth, which formerly was located in the center of the toll plaza to the south of the toll gates, was removed and a new booth constructed as an addition to the present administration building.

Together with this change a new procedure was established whereby commutation ticket sales are now made by toll collectors in the lanes without delay to purchaser. Formerly sales were made at the Sergeant's booth, resulting in considerable delay to purchasers and also congestion of the traffic lanes by vehicles being parked at the booth waiting to purchase tickets.

### Maintenance

The most costly item of maintenance for the world's longest single span suspension bridge is painting.

Experiments have been continuously conducted with sample of new metal protective coating materials to test rust-resisting qualities in comparison with the paint now being used.

During the last year, a total of over 28,000 tons of structural steel and wire cables was painted. They consisted of the San Francisco approach spans, arch span, main cables, and suspender ropes and assemblies, including eye bars, which provided excellent protection to the structure and cables.

### Engineering

The special study, initiated during the previous year, into the matter of providing a permanent protective fender around the North (Marin) Pier has been completed.

During the planning stages of the Golden Gate Bridge project, a ringwall for protection of the North pier and underlying rock exposures from extreme exposure to wave and current actions was discussed. But at the time such a wall was not considered necessary or desirable by the chief engineer, Joseph B. Strauss.

They considered permanent protection of the pier site as advisable but not necessary at this time. After thorough consideration of the reports and frequent questioning of the engineers, the bridge board, on the recommendation of its joint Finance and Building and Operating Committees, decided to defer action on the matter.

The results of other engineering surveys and inspection of the bridge structure, approaches and facilities showed all to be in excellent condition.

### Personnel

The bridge district personnel was a vital force in the bridge performance for 1949-50. The skill, vigilance and loyalty of our employees contributed immeasurably to the success of bridge operations.

All full-time employees are member of the California State Employees Retirement System under a cooperative arrangement by which both the district and the employees contribute to the cost of the retirement plan. The employees, all of whom work a 5-day, 40-hour week, were given a 15-days' annual vacation, equivalent to three weeks, by order of the directors, effective July 1, 1949. □

## SIXTY YEARS OF SERVICE

# Highlights from Annual Reports

## YEAR: 1950 — 1951

T here was another Foreword which declared, "THE GOLDEN GATE BRIDGE, linking San Francisco and the great Redwood Empire of California to the north, continues to pay its own way year after year from its total revenues, just as it has from the first day it was opened to traffic on May 27, 1937.

"But even if it were not a self-liquidating project, the bridge would be worth all its cost and more in the tremendous impetus it has given to the physical unification of the San Francisco Bay area; and to the growth and prosperity of San Francisco and its environs.

"The Golden Gate Bridge, aside from its revenue producing capacity, is an asset of incalculable value to San Francisco, California and the nation."

*[Editor: Pride of accomplishment? More than that. The arguments of the early days, the pursuit of a modest amount of money to build "the bridge that couldn't be built" and the spirit of the people who fought so hard for it, are now self-evident.*

*The Bridge is no longer an idea or a dream; it is concrete, living proof that the right thing was done at the right time. Without the Bridge and its pioneers and its backers and its early Boards and Administrators, the engineers and, yes, the painters, too, the future of Northern California would not have happened the way it did.*

*The Foreword of this Report showed the truth of the dream by pointing out the growth of the counties for the decade, 1940-1950 increased by 51.6%.*

*The modest footnote said, "Indications are that California's remarkable gain in population in the last decade will lose little of its momentum in the next ten years."*

*Truer words were never spoken.]*

### Report of the President

Traffic for all classifications showed an increase of 8.37 per cent last year. The largest gains were shown by transient or casual passenger automobiles and auto trailers, which increased 14.06 percent and trucks, which increased 14.5 per cent.

The unusually strong and stable financial position of the bridge has been maintained this last year. The district continues to pay out of revenues all expenses of operation and maintenance and insurance, including interest on outstanding bonds, and still have a substantial balance for the bond amortization and insurance reserve funds.

### Marin County Approach Improvement

After exhaustive surveys and studies, the board, in 1947, recommended construction of a new approach which would be at a 3.3 per cent grade maximum, to the east of the present Waldo approach. Estimated cost of this new approach was $12,000,000 at the time.

*[The Report cover for this year was a duo-tone (two colors) and featured a front-page logo circle on the front page, along with a Bridge photo which bore the legend inside: "THE COVER ◇ The Golden Gate Bridge is shown in all its glorious beauty from the shores of the Presidio military reservation in San Francisco. From this vantage point and from the area around old abandoned Fort Point in the left foreground, visitors and tourists come to try their luck at fishing in San Francisco Bay just inside the Golden Gate. In the background are Marin County shoreline and hills."]*

---

*In the rest of the World . . . .*

## HISTORICAL SNAPSHOT

**Events.** *North Korean and Chinese forces take Seoul, later pushed back to 38th parallel. Truman relieves MacArthur of command in Korea. Juan Peron re-elected president of Argentina.*

**Music.** *Gian Carlo Menotti, Amahl and the Night Visitors (opera); The King and I, Paint Your Wagon, on Broadway. Fender produces first electric bass guitar.*

**Books.** *James Jones, From Here to Eternity; Yasunari Kawabata, Thousand Cranes; J D Salinger, The Catcher in the Rye; Herman Wouk, The Caine Mutiny.*

**Films.** *The African Queen, An American in Paris, The Day the Earth Stood Still, Le Plaisir, Rashomon.*

**Television.** *CBS' "See It Now" features live, split-screen with Brooklyn Bridge and Golden Gate Bridge — to demonstrate television now coast-to-coast.*

**Everyday Life.** *Transcontinental dial telephone service available for first time.*

**The Golden Gate Bridge is recognized as a dynamic economic force in the development of the North Bay.**

# 1951

Previously in 1945, the state legislature had enacted a law giving the district the right to borrow $5,000,000 from the State Bond Retirement Fund of 1943 for the purpose of constructing an approach. However, every effort to obtain additional financing or an outright appropriate from state highway funds was unavailing.

The state department of public works and the state division of highways consistently advocated the widening of the Waldo approach to six lanes, divided, with construction of a new four lane tunnel near the summit, in addition to an existing four lane tunnel.

Execution of a contract with the state to carry out the work has been authorized by the bridge board, and it is expected that the project can be completed within two years. Any moneys in excess of $5,000,000, which may be required for the work, will be appropriated by the state highway commission.

## Proposed San Rafael-Richmond Bridge

During the last year, construction of a state-owned toll bridge between Richmond, Contra Costa County, and San Rafael, Marin County, has been recommended by the state department of public works. The location recommended was on a line approximately parallel to the existing vehicular ferry between those two points. Estimated cost of the crossing at this time is $54,000,000.

The bridge district directed its traffic engineer to make a study of what effect, if any, the proposed new bridge would have on traffic on the Golden Gate Bridge.

It was estimated that at 1950 levels the traffic which might be diverted from the Golden Gate Bridge to the new bridge, would amount to 622,000 passenger cars and 42,000 trucks. This represents 6.9 per cent of the annual volume.

## Report of the General Manager

The Golden Gate Bridge continues to keep pace with the growth and development of San Francisco and the counties of California to the north and south which it serves.

A total of 10,110,746 crossed the bridge during the year, while total operating revenue was $3,985,673. Operating costs were kept low with expenditures totaling $871,306.

## Traffic

Tolls were reduced with the beginning of the 1950-51 fiscal year. The principal reductions were from 50 cents to 40 cents as a one-way toll from passenger automobiles and adoption of a rate for trucks based upon the number of axles, instead of weight.

Truck traffic increased for the year by 14.5 per cent, being a increase of 90,639 vehicles. Truck revenue for the year decreased $146,498 or 20.7 per cent.

Bus traffic declined by 7.8 per cent. This decrease is a continuation of the downward trend that has been evident during the past three years despite the substantial increase in population of the area in Marin County and to the north.

## Maintenance

Wind instruments and seismographs record continuous bridge action, and accurate information is obtained and compiled as to the various movements caused under different wind and storm conditions. No detrimental results have been recorded to date.

While some harmonic motion in the bridge was revealed in the eight-year-period, the amount was only a small fraction of the movement that was provided for in the design of the bridge.

An experienced painting crew of 30 men is steadily engaged to prevent deterioration of the exposed steel surfaces, using paint prepared in accordance with a special formula for the Golden Gate Bridge exclusively.

Plans and specifications have been ordered prepared for construction of new traveling painters' scaffolds to hang under the bridge floor and the distribution trusses for the suspended spans. The work is being done by Clifford E. Paine, who was the principal assistant engineer under Chief Engineer Joseph B. Strauss during the construction of the bridge.

Improvements were made in the toll lanes at the Toll Plaza to speed the flow of traffic through the installation of new illuminated directional arrows and guide signs for motorists.

## North Pier Protection

The general manager has recommended construction of a circular or elliptical concrete fender around the North Pier similar to the fender build around the South Pier at the time the bridge was built. The matter is pending before the joint finance and building and operating committees of the board of directors.

## New Warehouse-Garage Building

Construction of a new building and equipment for a warehouse-garage and machine shop on land adjoining the Toll Plaza to the west, is again recommended. Present facilities are entirely inadequate. It is hoped that this construction will be authorized and can proceed as soon as the restrictions on critical building materials, in effect during the present national emergency, are relaxed.

## San Francisco Approach

With traffic increasing over the Golden Gate Bridge, it has become apparent that additional lanes are necessary to handle the flow of traffic heading north on the approach to the toll lanes.

The general manager has recommended and the board of directors authorized a request to the federal government for additional land on the east side of the San Francisco approach so as to permit widening for two additional lanes of traffic.

It is hoped that this widening can be carried out in the spring of next year, provided necessary materials are available.

*[The "national emergency" mentioned above and the concern for the availability of materials were references to the U.S. involvement in defending South Korea. The so-called Korean Conflict created shortage of some strategic goods, including building materials, especially steel and copper. — Ed.]*

## Legislation

The 1951 session of the state legislature considered many bills, directly or indirectly affecting the bridge district, which maintains its representative in Sacramento during the legislative session.

In addition, a special states senate committee was created to study problems concerning the inclusion of the Golden Gate Bridge into the state highway system. The bridge directorate cooperated in every way with the committee during its studies. □

**Fine Tuning.** *Even before the Golden Gate Bridge opened, the job of maintaining it began. Inspectors often "walk" the main cables, checking for potential problems. Here a tethered worker adjusts the turnbuckle that maintains tension on the hand-rope guides that assist maintenance cable-walkers.*

# SIXTY YEARS OF SERVICE
# Highlights from Annual Reports
## YEAR: 1951 — 1952

### Report of the President

This year marks an important anniversary for the Golden Gate Bridge — its 15th year in business.

But the fabric of events that is our 15 year history holds more than just the story of a bridge.

It is the story of 15 years of successful operation; of financial achievement; of service to the people; and of helping build California.

It is the story of a great self-liquidating project which has not only given tremendous impetus to the establishment of the beginnings of a genuine metropolitan area around San Francisco bay, but which has also contributed immeasurably to the growth and prosperity of San Francisco and the great Redwood Empire to the north.

And 1951-52 was another year of traffic growth, which, coupled with economical and efficient operation, added to the financial stability of the bridge district.

The Golden Gate Bridge is being operated as a public trust by a Board of 14 directors, representing the six counties of the bridge district. By state law under which the district was formed and incorporated December 4, 1928, the board is the governing body of the district and is charged with outlining the rules of policy.

As pointed out in previous annual reports, there never has been a deficit since the bridge was opened and it continues to pay its own way without tax subsidy of any kind from the start.

### Traffic Growth

Traffic has increased progressively every year on the Golden Gate Bridge since the termination of hostilities in Europe in 1945, and last year saw traffic reaching a record peak of 11,195,581 vehicles.

If the future traffic growth continues at the same rate experienced during the last seven years, total traffic will be doubled by 1960 to some 22,000,000 vehicles.

Should such volume actually materialize, a real problem would result for the present Golden Gate Bridge structure of six traffic lanes would not be able to accommodate it.

The day is not too distant when a second Golden Gate Bridge or another crossing between San Francisco and Marin county of some type is needed. Such a study as to location and when needed has already been proposed although no definite action has as yet been taken by the bridge district.

[In many ways the 15th anniversary of the Bridge was the most dramatic in its history. A severe winter storm caused the first closure of the span since it opened in 1937. Closing what was then the world's longest suspension bridge, even for a short period of time, was of concern to the District, whose members could vividly recall the details of one of bridge building's darkest hours — collapse of The Tacoma Narrows Bridge in 1940. The story of that tragedy was on the minds of everyone in December 1951. Highlights of that story are on page 191.

The Report begins mildly enough with the usual facts and figures, but contains a vivid, detailed account of a storm-tossed night.]

---

*In the rest of the World . . . .*
## HISTORICAL SNAPSHOT

**Events.** *Dwight D. Eisenhower elected President; Richard Nixon, Vice President. Japanese Peace Treaty approved, Japan regains sovereignty and independence. Eva Peron dies of cancer in Argentina. First sex-change operation performed in Denmark.*

**Books.** *Ralph Ellison, The Invisible Man; Hemingway, The Old Man and the Sea; Steinbeck, East of Eden.*

**Drama.** *George Axelrod, The Seven Year Itch; Truman Capote, The Grass Harp; Agatha Christie, The Mousetrap.*

**Films.** *Breaking the Sound Barrier, B'wana Devil (first 3-D film); Greatest Show on Earth, High Noon, Ikiru, The Quiet Man; Singin' in the Rain; debut of, This is Cinerama.*

**Everyday Life.** *Panty raids. Holiday Inn founded in Memphis. Sony markets the first pocket-sized all-transistor radio.*

## 1952

## Engineering Studies

Meanwhile the safety and security of the present Golden Gate Bridge must be guaranteed for all time, since it is the economic life-line between San Francisco and the productive and scenic northern coastal area.

It was for this reason that the Board of Directors decided to engage the best engineering brains in the country to study and appraise the whole bridge structure after a violent windstorm swept the bridge on December 1, 1951.

Under the directors' authorization, the District entered into contracts with three eminent consulting engineers — Messrs. Clifford E. Paine of Chicago, O.H. Amman of New York and Charles E. Andrew of Seattle — to study the problem of aerodynamic stability of the bridge. Mr. Paine is chairman.

The engineering consultants will not complete their study until early in 1953 because of wind tunnel tests and other detailed technical work of a complicated nature to be carried out.

Mr. Paine pronounced the Golden Gate Bridge as sound as ever after supervising the repair of minor damage caused by last December's storm. However, he felt that much could be done to the bridge to make it ride smoother in future wind storms, particularly through the addition of a bottom lateral bracing system which is now being studied.

Pending the out come of these engineering studies, the directors have deemed it both advisable and necessary to delay consideration of several major projects and betterments.

These include widening of the San Francisco approach; the construction of a much-needed warehouse, garage and machine shop; and likewise the installation of new safety scaffolds for painters and other maintenance workers on the bridge proper.

## Marin County Approach

Plans, now being drafted by the State Division of Highways, are expected to be completed and submitted early in 1953 to the Bridge Board for approval, as provided in the contract with the state. The work will then be advertised for construction bids and completion of the project within two years is anticipated.

## Report of the General Manager

During the year, the one hundredth-million motor vehicle rolled across the Golden Gate Bridge and confounded once again all the experts who try to size up the California of tomorrow by looking at the California of today.

And, in its fifteen years of operation, the Golden Gate Bridge, like California, has kept growing just too fast with traffic to conform to any of the formulas the experts use for guessing at next year's needs.

In its 15th year, the Golden Gate Bridge total traffic reached an all-time peak of 11,195,581 vehicles for 1951-52, an increase of 1,086,410 over the previous year or 10.7 per cent.

## Revenue

Total operating revenue for 1951-52 amounted to $4,251,913, an increase of 6.7 per cent over 1950-51.

## Traffic

Automobile traffic this year increased 11.5 per cent. Certain unusual conditions were experienced during the year which tended to distort the relationship of automobile traffic from normal.

During the period March 1, 1952 to and including May 20th, Pacific Greyhound Lines discontinued bus service between San Francisco and Marin County due to a strike of bus drivers. This period of two and one half months forced a change in the travel habits of many Marin County residents, shifting them to private automobiles.

## Storm Damages and Repairs

A violent windstorm occurred in the San Francisco Bay Area on December 1, 1951. These winds were of such intensity and character at the site of the Golden Gate Bridge that they caused vibrational motion in the suspension spans of greater magnitude than any previously recorded. However, higher winds of shorter duration had been experienced in the past.

Because the wind, poor visibility, and the movement of the span made driving across the Bridge both difficult and hazardous, the Bridge was ordered closed to traffic by the General Manager from 5:55 p.m. to 8:45 p.m.

**3:05 p.m.,** the wind velocity reached 50 miles per hour with frequent gusts at 55 miles per hour.

**4:30 to 5:30 p.m.**, the velocity in general remained at about 50 miles per hour, but gusts increased to 60 and 65 miles per hour.

**5:30 to 6:00 p.m.**, the general velocity increased to 55 miles per hour, and more frequent gusts ranged from 65 to 69 miles per hour.

**5:55 p.m.**, the peak recorded was 69 miles per hour. From that time on, the wind velocity steadily diminished so that at 7:00, the general wind velocity was reduced to a little more than 40 miles per hour with gusts to 50 miles per hour or less.

**9:00 p.m.**, the wind velocity ranged between 30 and 45 miles per hour, and continued to diminish.

The wind direction at 3:00 p.m. was Southwest. At 3:30 it became about twenty-two and one-half degrees south of west and held this direction quite constantly until 7:00 p.m. when it began to work a little more westerly. From 8:00 to 9:00 it was west.

No measurement was made of the lateral (side-to-side) deflection of the Bridge, but ten vibration recording instruments charted the vertical movement.

The charts showed that the main span was vibrating asymmetrically — the north segment moved upward while the south segment moved downward and vice versa.

In the longitudinal direction of the Bridge, the center span moved the full amount of travel which the construction permitted, which was 18 inches. The mid-point of the center span cabled moved beyond those limits at least 27 inches in each direction.

The Bridge, of course, was designed to move under strain and stress. Provision in the design was made for maximum transverse deflection at the center span of 27.7 feet; maximum downward deflection, center span, 10.8 feet; maximum upward deflection, center span 5.8 feet; transverse deflection of the towers, 12 1/2 inches; and longitudinal deflection of the towers, shoreward — 22 inches and channelward, 18 inches.

Engineering inspection of the suspension structure from anchorage to anchorage disclosed only superficial injury except at the expansion joints where the lateral system of the center span connects with the towers.

*(Continued on page 192.)*

## The Tacoma Narrows Bridge

The original 5,939-foot-long Tacoma Narrows Bridge opened to traffic on July 1, 1940, after two years of construction. The bridge, linking Tacoma and Gig Harbor in Washington state, collapsed just four months later during a 42 mile-per-hour wind storm on November 7.

The bridge earned the nickname, "Galloping Gertie" from its rolling, undulating behavior. Motorists crossing the 2,800-foot center span likened the trip to a ride on a giant roller coaster as they watched cars ahead disappear completely as if they had dropped into the trough of a large wave.

The original bridge was a suspended plate girder type that caught the wind, rather than allowing it to pass through. As the wind's intensity increased, so did the bridge's rolling, cork-screwing motion — until it finally tore the bridge apart.

An eyewitness account provided by Leonard Coatsworth, a Tacoma newspaper editor:

"Just as I drove past the towers, the bridge began to sway violently from side to side. Before I realized it, the tilt became so violent that I lost control of the car. I jammed on the brakes and got out, only to be thrown onto my face against the curb.

"Around me I could hear concrete cracking. I started to get my dog, Tubby, but was thrown again before I could reach the car. The car itself began to slide from side to side on the roadway.

"On hand and knees most of the time, I crawled 500 yards or more to the towers. My breath was coming in gasps; my knees were raw and bleeding, my hands bruised and swollen from gripping the concrete curb. Toward the last, I risked rising to my feet and running a few yards at a time. Safely back on the toll plaza, I saw the bridge in its final collapse and saw my car plunge into the Narrow."

It would be ten years before another crossing was built. After 29 months of construction, a newer, safer Tacoma Narrows Bridge opened on October 14, 1950. The new bridge spans 5,979 feet, forty feet longer than the original.

The sunken remains of "Galloping Gertie" were placed on the National Register of Historic Places in 1992 to protect her from salvagers. □

The 1951 storm that closed the Bridge brought to mind the death of a suspension bridge caused by high winds more than a decade earlier in pre-war, pre-television days. America saw the demise of "Galloping Gertie" in newsreels when they attended movies. In 1951, the high-wind closure brought those newsreels alive again on b&w television.

At the latter points the sliding block joints were damaged so that replacement of one bearing plate and the block was necessary at the San Francisco Tower while at the Marin Tower replacement of both bearing plates and the block was required. Total cost of the repairs was $72,880.

After completion of the repairs, the bridge was declared by Consulting Engineer Clifford E. Paine as sound as ever.

For the first time since its construction, the character, intensity and direction of the wind resulted in movements on the bridge floor of important magnitude.

Because Mr. Paine believed that much could be done to improve this behavior, comprehensive engineering studies were authorized by the Board of Directors to be carried out throughout 1952.

### Sausalito Lateral Slide

Continuous rains — which made the seasonal rainfall the heaviest in 71 years in the San Francisco bay area — seriously undermined the land surrounding the northerly terminal of the Sausalito Lateral approach to the bridge in Marin county near the Sausalito city limits, causing collapse of the wooden bulkhead at this point on January 24, 1952. In addition, the roadway pavement cracked for a length of almost 200 feet and dropped in some portions as much as 12 inches at first. Additional slides occurred later.

It was necessary to close the lateral completely to all traffic from January 31 to February 6 while slide material was removed.

### Maintenance

Painting, the largest single item of maintenance, was centered on the Marin or North Tower, the 317-foot arch span and the three San Francisco truss spans.

Bids were called for construction of new traveling painters' scaffolds to hang under the bridge floor. However, because of excessive costs, the bids were rejected and plans held in abeyance. ☐

SIXTY YEARS OF SERVICE

# Highlights from Annual Reports

YEAR: 1952 — 1953

## Report of the President

Noteworthy gains were made by the Golden Gate Bridge and highway district in the fiscal year ending June 30th, 1953, and it is gratifying to report that the District continues to maintain its sound and stable financial position.

As California's great post-war growth showed no signs of abating — the state's population increases 1,000 a day — rising motor traffic on the highways reflected the trend, and traffic over the Golden Gate Bridge responded accordingly.

## Tolls

The average automobile rate for total automobiles cross the bridge has gradually reduced from 50.6 cents per vehicle at bridge opening in 1937 to 34.7 cents for 1952-53, and it is estimated that the rates will be 33.8 cents for 1953-54.

## Legislation

Several bills affecting the Golden Gate Bridge and Highway District were introduced at the 1953 session of the California State Legislature. The major bills dealt with state acquisition and operation of the Golden Gate Bridge while another called for a mandatory 25-cent toll for passenger automobiles on the bridge for a one year trial period.

These major measures became highly controversial with the Board of Directors and a majority of the Boards of Supervisors of the counties within the Bridge District actively opposing them as inimical to the best interests of the District as a whole, the taxpayers, bondholders and bridge users.

The bills for state control of the bridge were defeated but the 25-cent toll bill was passed by the legislature and sent to the governor for approval on June 8, 1953.

Governor Earl Warren withheld his signature from (the Bill) on July 10, 1953 and thus it failed to become a law.

## Bridge Structural Changes

Plans and specifications for installation of a bottom lateral system of bracing on the Golden Gate Bridge were authorized on January 30, 1953 on the basis of an intensive study and report by a board of three eminent engineers, headed by Clifford E. Paine.

*[Editors' Note: An excerpt of Paine's report follows the General Manager's Report in this summary.]*

## Engineer for District

Mr. Clifford E. Paine was appointed by the Board of Directors as Engineer for the District, effective May 1, 1953 to succeed Mr. John G. Little, who had served as consulting engineer since 1947, and contributed to help insure the success of the bridge operations.

---

*In the rest of the World . . . .*

## HISTORICAL SNAPSHOT

**Events.** *Korean armistice signed at Panmunjom. Earl Warren appointed Chief Justice of U.S. Supreme Court. Convicted spies Julius and Ethel Rosenberg executed. Edmond Hillary and Norkey Tenzing reach Mt. Everest summit. Discovery of the double helix structure of DNA.*

**Music.** *Benjamin Britten, Gloriana; Shostakovich, Symphony No. 10; Kismet & Wonderful Town open on Broadway. Marty Robbins hit, Singin' the Blues.*

**Books.** *Ray Bradbury, Fahrenheit 451; Raymond Chandler, The Long Goodbye; Ian Fleming, Casino Royale (the first James Bond).*

**Deaths.** *Joseph Stalin, Prokofiev, Hilaire Belloc, Dylan Thomas, Eugene O'Neill.*

**Films.** *The Robe (first film in Cinemascope); Roman Holiday, Shane, dir George Stevens; Monsieur Hulot's Holiday, dir/star, Jacque Tati; Tokyo Story, dir Yasujiro Ozu.*

**Everyday Life.** *(Well, not quite.) John Fitzgerald Kennedy marries Jacqueline Lee Bouvier in Newport, Rhode Island.*

## 1953

Mr. Paine was associated with Mr. Joseph B. Strauss, chief engineer for the Golden Gate Bridge throughout its construction, and was principal assistant bridge engineer during which he supervised the work in the field and shops to insure its execution in accordance with the intent of the bridge design.

Now, returning as engineer for the District, he will supervise the installation of the bottom bracing on the bridge.

### Report of the General Manager

Total traffic for the year again set a record as 11,723,318 vehicles crossed the bridge, and increase of the prior year of 4.3%. Although this was a new peak, the per cent of increase was the lowest experienced during the past eight years.

### San Francisco Approach and Exit Widening

Traffic checks and observation of peak hours of vehicles into the toll plaza indicates the need for widening the westerly roadway between the toll plaza and the Funston Avenue turnoff so as to provide for a more free flow away from the toll gates.

On a Sunday, 78.1% of total south bound bridge traffic during the evening peak hours turns off into Funston Avenue.

### Maintenance

A total of nine thousand tons of steel was painted during the past year requiring 7,900 gallons of paint to protect the metal surface of the bridge.

So as to make certain the best quality paint for bridge use is employed, more than eight hundred paint materials have been tested since 1936, at a cost of $25,000, in search of paint protective coatings of greater durability and wear. Paint tests are carried on constantly at the bridge site on exposed metal test panels, as all paints used on the structure must withstand the heavy fog, rain, wind, salt water spray and sunshine which prevail at the Golden Gate.

Since increased bridge traffic demands that parking of District's repair equipment on the roadway be held to a minimum to aid in preventing accidents, the air-powered staging and winches, operating from the sidewalk level, remove this hazard and provide a safe and fast means for hoisting maintenance workmen to and from tower and cable tops without interfering with traffic.

High above the deck of the bridge roadway, the suspender rope separator assemblies are located, where U-bolt attachments guide and hold the long cables apart, preventing wind action from bumping the cables against each other. With new airpowered swing cages, riggers are now able to inspect and replace the U-bolts in a minimum of time. Four hundred and eighty new U-bolts were installed within the last three months, replacing sixteen-year-old units.

At the end of 1952 the Bridge Management found that certain changes would be necessary to modernize the present sixteen-year-old toll collection equipment to reduce maintenance costs and speed up traffic as well as collections. As a result, new toll collections have been ordered for the two truck lanes which will be ready for next year's business.

### Bridge Alterations

Comprehensive engineering studies on the Golden Gate Bridge, which were authorized by the Board of Directors as a result of noticeable motions that occurred in a severe windstorm on December 1, 1951 were completed.

Work is scheduled to begin in the early fall of 1953, and be completed within 320 days after signing the contract. It is not expected to interfere with the normal flow of traffic on the bridge.

Excerpt from:
"REPORT ON
ALTERATIONS OF
GOLDEN GATE
BRIDGE"

By Board of Engineers
Clifford E. Paine,
Chairman
Othmar H. Ammann
Charles E. Andrew

## Summary of Conclusions and Recommendations

As a result of the exceptional storm of December 1, 1951 the Golden Gate Bridge was subjected to the severest oscillating motions since its completion in 1937.

Available knowledge derived from the behavior of similar structures and from extensive research and experimental work which has been carried on in this country and elsewhere since the failure of the Tacoma Narrows Bridge in 1940, supplemented by the studies and model tests made as part of our investigations, lead us to conclude that the steadying of the Golden Gate Bridge to prevent future objectionable motions and to secure an ample margin of safety against dynamic wind action is entirely practicable.

The motion which we are most concerned about is the torsional first asymmetric one in which the main span oscillates with a twisting motion. We believe that the recommended improvements will eliminate this motion within the range of wind velocities likely to occur.

As a first step we recommend the addition of a system of bottom laterals in the plane of the bottom chords of the stiffening trusses for the entire length of the suspension span.

We believe this first step will so materially improve the behavior of the bridge under wind action that no further improvements may become necessary.

Our detailed studies indicate that the installation of such a system of bottom laterals is within the safe carrying capacity of the bridge and can be carried out without interference with the flow of traffic over the bridge.

This installation is estimated to involve a total expenditure of approximately $3,500,000. Its completion will require from nine to twelve months time.

As a second step to be undertaken in case the first step should prove not entirely adequate we recommend installation of a system of stays connecting the cables at the center of the main span with the floor structure.

This step would in our opinion materially enhance resistance against asymmetric torsional motions (the motions which ultimately broke the floor-structure of the Tacoma Narrows Bridge) and otherwise help the steadying of the bridge.

A third step which, according to the results of our model tests, would assist in steadying the bridge, consists of a modification of the footwalk structure.

The existing solid footwalk slabs would be replaced for part or the whole of their width by open steel gratings. This improvement would be very costly, but would be entire practicable without serious interference with vehicular traffic over the bridge.

We recommend that detailed plans and specifications for the bottom lateral system be prepared immediately, that thereafter competitive bids be invited and that construction be expedited as speedily as possible. □

# The Golden Gate Bridge
## "Numbers" Page

### BRIDGE

| | | | |
|---|---|---|---|
| Total Length of Bridge including approaches | 1.7 mi. | 8,981 ft | 2,737 m |
| Length of suspension span incl main & side spans | 1.2 mi. | 6,450 ft | 1,966 m |
| Length of main span | | 4,200 ft | 1,280 m |
| Length of side span | | 1,125 ft | 343 m |
| Width of Bridge | | 90 ft | 27 m |
| Width of roadway between curbs | | 62 ft | 19 m |
| Width of sidewalk | | 10 ft | 3 m |
| Clearance above mean high water | | 220 ft | 67 m |
| Deepest foundation below mean low water | | 110 ft | 34 m |
| Total weight of Bridge, anchorages and north and south approaches (1937) | | 894,000 tons | 811,500,000 kg |
| Total weight of Bridge, anchorages and north and south approaches (1994) | | 887,000 tons* | 804,700,000 kg* |
| Weight of Bridge not including anchorages and north and south approaches and including suspended structure, towers, piers and fenders, bottom lateral system and orthotropic redecking (1994) | | 419,000 tons | 380,000,000 kg |

### TOWERS

| | | |
|---|---|---|
| Height of towers above water | 746 ft | 227 m |
| Height, towers above roadway | 500 ft | 152 m |
| Base dimension (each leg) | 33 x 54 ft | 10 x 16 m |
| Load on tower from cables | 61,500 tons | 56,000,000 kg |
| Weight of the two towers | 44,4000 tons | 40,200,000 kg |

### MAIN CABLES

| | | |
|---|---|---|
| Diameter, over wrapping | 36 3/8 in | .92 m |
| Length of one cable | 7,650 ft | 2,332 m |
| Total length of wire used | 80,000 | 129,000 km |
| Number of wires each cable | 27,572 | |
| Number of strands each cable | 61 | |
| Weight of Main Cables, Suspender Cables & Accessories | 24,500 tons | 22,200,000 kg |

*The total weight shown for 1994 accounts for the reduction in weight due to replaced the deck between 1982 and 1986. The weight of the original reinforced concrete deck and its supports strings was 166,397 tons (150,952,000 kg). The weight of the new orthotropic steel plate deck, its two inches of epoxy asphalt surfacing, and its supporting pedestals is now 154,093 tons (139,790,700 kg.) This represents a total reduction in weight of 12,300 tons (11,158,400 kg), or 1.37 tons (1133 kg) per lineal foot of deck.

## SIXTY YEARS OF SERVICE

# Highlights from Annual Reports

## YEAR: 1953 — 1954

### Report of the President

Financial stability of the Golden Gate Bridge and Highway District was emphasized in the result of operations during the 17th fiscal year ended June 30, 1954.

### Tolls

Tolls on the Golden Gate Bridge are constantly under study by the District's Finance Committee and during the past year consideration was given to various proposals for toll adjustments.

The Board agreed with the Finance Committee that present tolls are fair and equitable, and as low as possible in view of financial obligations of the District, and that further reduction in bridge revenues was unwarranted at this time.

### Insurance

Cost of replacement of the Bridge and facilities in the event of disaster, is estimated to be in excess of $100,000,000.

### Looking Forward

As the Bridge District enters its eighteenth year of operation, there is every indication that, barring the unforeseen, traffic over the Golden Gate Bridge will continue to increase in keeping pace with the growth of California and soaring motor vehicle registrations.

### Report of the General Manager

Traffic volume again set a new record, as it has each year since 1945. The trend of traffic increase however, has leveled off to a slope consistent with the population growth of the North Bay.

Total vehicles cross the bridge during 1953-54 were 12,213,455, and increase of 4.62%. Bridge traffic in the last nine years has increased nearly 84%. Total Bridge Toll revenue was $4,502,861.

Analysis of traffic and revenue for the full year discloses interesting information. For the full year, automobiles made up 89.9% of the total vehicles and produced approximately 83% of total revenue. The interesting feature of the automobile traffic is that although there are reduced rate tickets available, the preponderance of automobile traffic passes through the bridge at cash toll of 40c.

This in spite of the fact that there is available a 40-trip 45 day commutation ticket selling at $8, which gives a cost per trip of only 20c, a 50% reduction from the cash toll.

The average toll per automobile for the year for all classifications was 34.07c. The average toll per revenue truck was 76.61c.

### Maintenance

Good maintenance combines the skill of men, the services of tools, and proper use of material to accomplish the best results. It has been management's policy to encourage such procedure, and to reach this goal, the District's forces have made remarkable progress over the year.

---

*In the rest of the World . . . .*

## HISTORICAL SNAPSHOT

**Events.** *Racial segregation in U.S. public schools declared unconstitutional. First atomic-powered submarine, U.S.S. Nautilus, christened. U.S. and Japan sign mutual defense agreement. Filter-tip cigarettes gain popularity as smoking health risks studies published.*

**Music.** *Von Karajan becomes conductor of Berlin Philharmonic. First Newport Jazz festival held. "Rock Around the Clock" released.*

**Films.** *Bad Day at Black Rock, Les Diaboliques, On the Waterfront, The Seven Samurai, White Christmas.*

**Books.** *William Golding, Lord of the Flies; Francois Sagan, Bonjour Tristesse.*

**Drama.** *Tennessee Williams, Cat on a Hot Tin Roof.*

**Sports.** *Roger Bannister runs the mile in under four minutes.*

**Everyday Life.** *The "Cha-Cha" introduced in U.S. Marilyn Monroe marries Joe DiMaggio.*

# 1954

Modern bridge painting and maintenance demand modern methods and equipment to do an effective job at reasonable cost, especially while working at great heights and in dangerous position, which generally prevail when working on a suspension bridge.

To make it safer and faster for the workmen to perform their duties, new air powered swing staging was used successfully for replacing 480 separator "U" bolts which hold the main cable suspender ropes apart in strong winds. The separators are 350 feet above the roadway at the towers and reduce to a minimum of 160 feet on the side spans. The suspender rope cables, consisting of 512 sets, varying in length from 490 to 23 feet, were repainted during the year with the same type of equipment.

The painting crew also covered over 900 lineal feet of stiffening truss, including floor systems, curbs and railings, representing 3,300 tons, for the main and side suspended spans.

The Marin shore span structure, between the north abutment and Marin anchorage, including the supporting towers, was also repainted.

Toll collection equipment, originally installed in 1937, was replaced with the latest improved units for the two truck lanes, and it is planned to install other modernized toll units as rapidly as possible.

## Bridge Improvements

A contract was executed on July 14, 1953 with Judson Pacific-Murphy Corporation for the installation of a bottom lateral system of bracing on the bridge to improve its behavior under high wind action.

The addition of approximately 5,000 tons of structural steel to the bottom of the center span as well as the two suspended side spans for a total length of 6,450 feet is expected to reduce any asymmetric torsional motion in a high wind to a point where it will be neither objectionable nor detrimental to the structure.

The work, being carried out under the supervision of Clifford E. Paine, principal engineer during construction of the bridge and now engineer for the Bridge District, entails an overall expenditure of approximately $3,500,000.

With completion of the steel erection for the bracing, there will follow tightening of some 3,700 cable band bolts on the suspended spans.

New traveling scaffolds under the bridge for painters and maintenance workers will also be required as a result of the additional bracing. Mr. Paine was authorized to prepare plans and specifications for their construction, which he will also supervise.

## Waldo Approach Widening

As a cooperative enterprise of the Bridge District and the State, this project calls for widening the Waldo Approach for some three miles from an undivided four lanes to a divided six-lane freeway of modern standards. It includes construction of a new three-lane, one-way tunnel paralleling the existing Waldo Tunnel.

Low bidder for grading, tunnel construction and necessary viaducts, overpass and other structure was Guy F. Atkinson Co. of South San Francisco to whom a contract was awarded for $4,122,382.

Work commenced in September 1953 and completion of the contract is expected by the end of 1954.

During the year a cooperative agreement was executed between the Bridge District and the State of California for the planning and engineer supervision of widening both sides of the San Francisco approach to the bridge between the Toll Plaza and the Funston Avenue cloverleaf, a distance of one-half mile.

This improvement is designed in the public safety to relieve traffic congestions approach and leaving the Toll Plaza at the south end.

Through the cooperation of military authorities, additional land of the Presidio military reservation of San Francisco was granted for the project by the government. □

SIXTY YEARS OF SERVICE

# Highlights from Annual Reports

YEAR: 1954 — 1955

## Report of the President

Accounts of the District show a net income for the year of $2,300,146.41, only slightly less than the prior year despite a reduction in the cash toll automobile rate from 40c to 30c on February 1, 1955. This net was after payment of $1,284,200 of interest due on outstanding bonds.

## Insurance

Geological and seismological studies were conducted by a board of eminent scientists at the request of the District to determine the present status of natural phenomena as related to the ability of the bridge to continue to withstand the forces of nature with the same degrees of safety already experienced over a period of eighteen years.

Findings of the Board of Engineers were favorable. It was stated in the report as follows:

"We conclude that scrutiny of areas adjoining both north and south foundations of the Bridge and the Bridge itself over a period of more than twenty years fails to reveal signs of surface instability.

"The observed behavior suggests that both foundation areas are essentially stable and no more likely to slide under the impact of future earthquakes than other areas of comparable relief in the Bay Area."

After these findings were received, a report was initiated for presentation to the insurance underwriters, seeking a more favorable premium rate and a high over-all coverage.

## Toll Rates

After careful and continuous study of the financial status of the District and the potential expenditures for structural improvements, the Board of Directors reduced the automobile cash toll rate from 40c to 30c, effective February 1, 1955, and discontinued the 20-trip $6.00 ticket.

## The Future

There appears to be every reason to expect that the populous development of this great State, with its industrial wealth, agricultural prominence and general growth, will insure the continued potential prosperity that will carry the trend of the past into the future. Barring unforeseen catastrophe, the only predictable barrier to the rapid rate of expansion in bridge traffic is the ability of the bridge itself to carry the volume under its limited physical dimensions.

## Report of the General Manager

Traffic across the bridge during the year totaled 13,220,641 vehicles, the highest on record. For the full year there was an increase of 7.8% in total traffic. During the year truck traffic increased 11.9%. This condition appears to reflect the accelerated development of Marin County and other counties to the north of San Francisco where population growth and residential expansion are widespread.

## Revenue

Total revenue for the year increased by only $6,579 over the prior year, or 0.15%. This was due to the reduction in automobile cash toll rate, which was in effect for five months of the year.

---

*In the rest of the World . . . .*

# HISTORICAL SNAPSHOT

**Events.** *Rosa Parks arrested in Montgomery, Alabama for sitting at the front of a bus. U.S. minimum wage inceased to $1/hr. Following heart attack by Pres. Eisenhower, stock market lost $14-Billion on Sep 26. The labor organizations AFL and CIO merged.*

**Music.** *"Damn Yankees" opens on Broadway. Pop song hits, "Sixteen Tons" by Ernie Ford; "Roll Over Beethoven," Chuck Berry.*

**Books.** *James Baldwin, Notes of a Native Son; Vladimir Nabakov, Lolita.*

**Drama.** *Beckett, Waiting for Godot; William Inge, Bus Stop.*

**Deaths.** *Albert Einstein, Thomas Mann, Maurice Utrillo.*

**Born.** *Bill Gates.*

**Sports.** *Top U.S. golf money- winner, Julius Boros, $65,121. Dodger beat Yankees 4 games to 3 in 52nd World Series.*

**Everyday Life.** *Davy Crockett mania sweeps U.S. Disneyland opens in Anaheim.*

# 1955

## Maintenance and Operations

Of total operating expense, the cost of toll collections made up 36% and cost of insurance accounted for 20%. Of total maintenance and repair cost, painting represented about 50%. Approximately 22,000 tons of steel were painted during the year, and 6,500 gallons of paint were used.

Increase in vehicular traffic has necessitated additional toll collectors in order to handle properly the peak volumes with a minimum of delay to free flow.

## Structural Improvements

A bottom lateral system of bracing on the bridge was completed and accepted by the District as of January 7, 1955, adding a much higher factor of safety against possible damage from high velocity wind action.

The bottom laterals have resulted in an increase of 35 times the torsional rigidity of the bridge floor structure and an increase in over-all torsional resistance of the entire suspended structuring, including deck and cables of 2.75 times the corresponding resistance without the bracing.

## Tightening Cable Band Bolts

As a related project, an unusual maintenance procedure — the tightening of some 3,700 cable band bolts — was carried out on the bridge this last year at a cost of $69,000.

These bands clamp around the main cables from which the bridge deck is suspended. To prevent the bands from slipping along the main cable due to the steep slope, the two half sections are held together by two sets of threaded bolts. Tightening the nuts on these bolts increases the gripping action on the main cables.

Although tension in the giant bolts — most of them 2 3/8 inches in diameter and nearly 2 1/2 feet long — is checked a periodical intervals, about every five years, this was the first time since the bridge was opened in 1937 that all of them have been tightened. It was also the first time that giant impact wrenches had been used for such work.

## Round House Renovation

During the year, a fire in the Round House Restaurant at the Bridge Toll Plaza resulted in complete renovation of the kitchen and the restaurant interior.

## Widening of San Francisco Approach

Widening of the San Francisco approach to the bridge between the Toll Plaza and the Funston Avenue Cloverleaf was brought to completion during the year. This project provided two extra lanes in each direction and was instrumental in materially expediting traffic flow during peak volume periods.

Realignment of the access road and revision of the public parking area to the east of the Toll Plaza were carried out as part of the project, which was done under the supervision of the State Division of Highways and financed by the District at a cost of approximately $400,000.

## Widening of Waldo Approach

Work progressed on schedule during the year in the widening of the Waldo approach to the bridge in Marin County, a cooperative project of the Bridge District and the State of California.

During the year, a contract was awarded for completion of the final stage of the work, including the surfacing of 3.7 miles of roadway, lighting and complete tiling of the new tunnel and partial tiling of the old tunnel.

## Machine Shop and Garage

Plans were approved and a contract awarded early in the year for the construction of a much needed garage and machine shop building on military property acquired from the United States Government immediately to the west of the Administration Building.

## Traveling Scaffolds

A contract was awarded for construction of new motorized traveling scaffolds to be used for painting and maintenance work on the under-deck portion of the bridge. The scaffolds will not be completed and installed until 1956.

## Bridge Personnel

The success of bridge operations is dependent to a great extent on the skill, vigilance and loyalty of Bridge District personnel. The management is proud of the employees and the cooperation and teamwork that have contributed to make the bridge record of performance outstanding. □

SIXTY YEARS OF SERVICE

# Highlights from Annual Reports

YEAR: 1955 — 1956

## Report of the President

New records of traffic volume have been attained in keeping with the phenomenal growth of the areas served by the bridge which, in turn, has been a potent influence in the population expansion and commercial developments of the counties to the north of San Francisco.

Probably the most significant action taken by the Board of Directors during the fiscal period was the reduction, effective October 1, 1955, of the cash automobile toll to 25 cents.

## Waldo Approach

March 20, 1956 marked the completion of this vitally important project, and on that day the Golden Gate Bridge freeway was dedicated to public use by appropriate ceremonies.

As one of the features of that project there was developed at the north bridgehead an outstanding vista point where a magnificent panoramic view of San Francisco and the bay area is obtained. When finally landscaped and beautified, this vista point may well become one of the most attractive points of interest not only for the visiting tourists, but also for the residents of the entire bay area and, in fact, Northern California.

## Other Achievements

Other achievements of lesser important, but nevertheless significant, are the completion of the San Francisco approach between the Toll Plaza and Funston Avenue, and a much-needed new garage-machine shop.

Following a thorough renovation, the Round House restaurant at the Toll Plaza was re-opened July 1, 1955.

Progress was also made during the year on the construction of new motorized traveling maintenance scaffolds, necessitated by the installation of the bottom lateral bracing system in 1954. The erection of these scaffolds, designed by our engineer, Clifford E. Paine, will completely minimize the danger of accidents to workmen as well as facilitating their work.

## Planning for the Future

Two decades of prosperity should not be permitted to obscure the possibility of leaner years to come. The virtual monopoly that has been enjoyed may not be enduring.

Already this year, a new bridge has been constructed by the State of California, spanning upper San Francisco Bay between Richmond and San Rafael, and another bay crossing between San Francisco and Tiburon in Marin County is under study by the State. In addition, there has been proposed a rail rapid transit system between San Francisco and Marin County as part of a vast bay area program.

We can, I believe, look forward with optimism to the future; but, to keep pace with this fast-moving calvacade, no time should be lost in initiating the necessary studies to predict as nearly as possible the most prudent course of action to deal with these imponderables as the future brings them into focus.

---

*In the rest of the World . . . .*

## HISTORICAL SNAPSHOT

**Events.** *Eisenhower re-elected by landslide. Two airliners collide over Grand Canyon, killing 128. Saulk polio vaccine sold on open market. First trans-Atlantic telephone cable begins operation. U.S. Supreme Court rules bus passenger segregation unconstitutional. In Japan, first nationwide system of testing pupils begins. "In God We Trust" adopted as U.S. national motto by act of Congress.*

**Music.** *My Fair Lady; Elvis, Don't Be Cruel, Love Me Tender; Doris Day, Que sera, sera.*

**Books.** *Allen Ginsberg, Howl.*

**Deaths.** *A A Milne, Bertolt Brecht, Alfred Kinsey, Bill Boeing.*

**Films.** *Giant, Early Spring, The King and I, Nuit et Brouillard, The Seventh Seal.*

**Everyday Life.** *(Again, not quite.) Grace Kelly marries Prince Rainier; Marilyn marries writer Arthur Miller.*

# 1956

### Report of the General Manager

For the year, total traffic reached 14,749,185 vehicles and total revenue from traffic and miscellaneous amounted to $4,112,357. Traffic increased by 1,528,544 vehicles or 11.6%. Due to the toll rate reduction, total revenue decreased by $405,247 or 9%.

Despite the continued growth of Marin County and other counties to the north, bus traffic declined during the year, indicating a diversion of passengers from the buses to automobiles.

### Toll Rates

As indicated above, toll rates for automobiles were again reduced during the year. Effective on October 1, 1955, the cash toll rate for automobiles was reduced from 30c to 25c and the commutation ticket rate was reduced from $8 to $7.

### Maintenance and Operation

The three largest items of expense for the year were Toll Collecting, $307,682; Painting, $201,459; and Insurance, $142,667, totaling $651,808, or 57% of total expenses of operation and maintenance.

## Improvements to Physical Properties

### San Francisco Approach Widening

On August 8, 1955, widening of the approach road on the San Francisco side of the toll plaza, was complete. This provided a much needed improvement by increasing the roadway width both northbound and southbound, by adding two traffic lanes in each direction.

### Waldo Approach Widening

On March 20, 1956, the newly widened Waldo Approach was officially dedicated and opened to traffic in both directions. Completion of this project marks one of the most important accomplishments during the 19 years of bridge operation.

Whereas the former roadway provided only two traffic lanes in each direction with no median division strip, had relatively short radius curves and comparatively steep gradient, the new project is designed with three lanes in each direction separated by a median strip and barrier, with longer radius curves and reduced gradient in certain sections.

### Other North Bay Improvements

Included in the 1956-57 State Highway Budget, as announced on October 24, 1955, were several improvement projects affecting the Golden Gate Bridge. Among these are the:

(1) Widening of San Francisco Approach to a divided eight-lane roadway from Richardson Avenue-Marina Ramp to the Funston Avenue cloverleaf.

(2) Extension of divided six-lane freeway in Marin County from Alto to Greenbrae and a grade separation at Corte Madera.

(3) Extension of five miles of divided four-lane highway into Santa Rosa, north of Petaluma, connecting with the eleven miles that are now under construction.

With the rapid rate of traffic increase on Highway U.S. 101 in Marin County and to the north, these improvements are vital as a means of expediting the free flow of traffic and relieving serious points of congestion that already exist.

### North Pier Soundings

New underwater soundings were taken in the spring by the U.S. Coast and Geodetic Survey in the vicinity of the Martin County North Tower supporting pier, to determine the extent of erosive action. This was the latest of a series of periodic Soundings at three-year intervals. In the report submitted by the U.S.C.G.S. it was stated, "The survey should allay any fears of serious erosion in the vicinity of the pier."

## Richmond—San Rafael Bridge

The new bridge is nearing completion and is expected to be opened to traffic in October 1956. At the outset it is expected that the new bridge will divert some traffic from the Golden Gate Bridge. However, at the rapid rate of north county growth this loss of traffic will be compensated by the normal increase on the Golden Gate Bridge, with no harmful effect upon District revenue. Actually, the overall effect will probably be to increase Golden Gate Bridge traffic during future years.

## San Francisco—Tiburon Bridge Study

Under legislative authority, consideration is being given to preliminary design of a second Marin County bridge extending from San Francisco to Tiburon via Angel Island. The study is being made by a Division of San Francisco Bay Toll Crossings of the California Department of Public Works. The bridge would be a two-level structure with provision for rail rapid transit on the lower deck. The southern terminus of the bridge would be near Aquatic Park in San Francisco between Larkin and Polk Streets.

## Rapid Transit Report

On January 5, 1956 the engineering firm of Parsons, Brinckerhoff, Hall and MacDonald submitted a report on Bay Area rapid transit to the San Francisco Bay Area Rapid Transit Commission. Among the major construction projects recommended in the report was the proposal that a rail rapid transit line be installed beneath the deck of the Golden Gate Bridge as a means of providing interurban passenger service between San Francisco and Marin County.

This proposal would entail major alteration in the structural members of the bridge and would add substantial weight to the suspension cables. The Golden Gate Bridge was not designed to carry rapid transit trains. The matter has been referred to Committee for consideration.

## Miscellaneous

On June 21, 1956, the 150 millionth vehicle crossed the bridge.

On December 18, 1955, wind velocity from the southeast reached 68 miles per hour during a heavy rain storm. The maximum double vertical amplitude as recorded on the vibration instruments, was only seven inches.

In comparison, during the storm of 1951, a wind of sustained velocity of 69 miles per hour caused a maximum vertical amplitude vibration of 130 inches causing the bridge to be closed to traffic. Since installation of the bottom lateral bracing there have been no objectionable vibrations of the bridge deck whatsoever. □

SIXTY YEARS OF SERVICE
# Highlights from Annual Reports
YEAR: 1956 — 1957

## THE 20th ANNIVERSARY

### Report of the President

Total vehicles for the 1956-57 fiscal year were almost one million greater than for the prior year, despite minor losses in auto and truck traffic to the new San Rafael–Richmond vehicular bridge, a state-owned and operated toll facility, which opened September 1, 1956.

Automobile traffic continues to make up the preponderance of toll paying vehicles, being 92.7 per cent for the year. Of total toll paying automobiles, commutation traffic contributed 25.1 per cent and cash toll autos 74.9 per cent.

### Richmond—San Rafael Bridge

The new bridge is a double deck structure with three traffic lanes on each level. Routing of the bridge is approximately the same as that of auto ferry boats that were replaced by the structure. The bridge connects the east shore of San Francisco Bay from a point in Richmond, Contra Costa County, to the west shore at a landing near point San Quentin, Marin County. Over-all the bridge is 21,343 in length with two cantilever spans at channel crossings.

Traffic during the first year of operations was estimated at 3,900,000 vehicle. Actual traffic the first twelve months, including the initial curiosity group, was only 2,588,320, or about 1,300,000 short of anticipation.

This deviation from estimate, although having no direct bearing upon the Golden Gate Bridge, should be seriously contemplated in making traffic projections into the future.

### Parallel Bridge Study

In the engineering report it was proposed that a six-lane suspension bridge be considered, with a provision for double decking at a later date and with design to include rapid transit. A main suspension span of 3,800 feet was suggested with side spans of 1,900 feet in length. Towers 615 high and a minimum vertical clearance of 220 feet for navigation were recommended. A second suspension bridge of 5,400 feet in length would span Raccoon Strait on the Marin County end.

It was concluded in the report that:

1. Such a bridge would be engineeringly feasible.

2. Additional information must be obtained before determining financial feasibility.

3. Present rate of traffic growth will require another bridge within seven to ten years.

4. Most satisfactory San Francisco terminus within statute limits would appear to be a point between Polk and Larkin Streets.

5. Alternate Route 101 along west side of Tiburon Peninsula appeared to be best suited to traffic and service condition.

The report said an additional $500,000 would be required to proceed with further necessary studies on financial feasibility, and another $1,000,000 to bring studies and estimates into condition for sale of bonds and commencement of construction.

---

*In the rest of the World . . . .*
## HISTORICAL SNAPSHOT

**Events.** *New transcontinental speed record set by Maj. John Glen in a Navy F8U-1P jet, 3 hours, 23 min., 8.4 sec. Troops sent to Little Rock to enforce segregation. Link between cigarette smoking and cancer established. World's longest suspension bridge, the Mackinac Straits Bridge open for traffic; cost: $100,000,000.*

**Books.** *Jack Kerouac, On the Road; Alan Watts, The way of Zen.*

**Music.** *Hindemith, Die Harmonie der Welt (opera); Elvis, All Shook Up; Buddy Holly, Peggy Sue; West Side Story.*

**Deaths.** *Bogart, Toscanini, Sibelius, Jos. McCarthy.*

**Television.** *New shows: Maverick, Perry Mason, Wagon Train. NHK in Japan broadcasts, "The True Face of Japan."*

**Film.** *The Bridge on the River Kwai, Gunfight at the OK Corral, Twelve Angry Men.*

# 1957

However, while the State Legislature passed a bill for an additional $100,000 for the project after receiving the preliminary report, it was disapproved by the Governor.

## Removal of Interurban Rail from San Francisco-Oakland Bay Bridge

Another study was submitted by the State Division of San Francisco Bay Toll Crossings, recommending removal of interurban rails from the San Francisco-Oakland Bay Bridge and repaving the area for automobile vehicles, together with other structural changes, at a total cost of $35,000,000. It was estimated that such alterations would increase the vehicular capacity of the bridge by from 25 to 35 per cent.

It is of interest to note this proposal to remove rails on the more heavily travelled Bay bridge, in view of the recommendation by other engineers that rails for rapid transit be installed on the Golden Gate Bridge where both present and potential traffic are much lighter.

## Bay Area Rapid Transit

During the year there was considerable activity on the subject of creating a San Francisco Bay Area Rapid Transit Authority and formulating appropriate legislation to provide the means of proceeding toward realization of a regional rapid transit program.

A bill was introduced at the 1957 legislative session. After much controversy and major revision, the bill was passed and approved by the Governor. During the coming year it is expected that the new legislation will be carried into effect by appointment of a Board of Directors and commencement of action on a rapid transit program.

## Improvement of Approach Highways to the Bridge

Much needed improvement on Highway U.S. 101 immediately north of the Golden Gate Bridge was realized during the year by completion of a new Richardson Bay bridge in Marin County. South-bound lanes were opened to traffic on September 20, 1956, and northbound lanes were opened on October 11, 1956.

A new high level concrete bridge was opened at the Greenbrae and Corte Madera Creek in Marin County. The new bridge handles southbound traffic and an old bridge, reconstructed, will provide three lanes for northbound traffic.

Progress has been made on construction of a double-deck Embarcadero Freeway in San

Francisco, which will ultimately connect with the Golden Gate Bridge. Construction was commenced during the year on a portion of the freeway between Harrison Street and Vallejo Street, passing in front of the Ferry Building.

## Structural Improvements

During the last year, it was determined the repairs were required to correct a condition on the south face of Pylon S-1 on the bridge, at the point where steel roadway stringers rest upon the concrete wall over the arch span.

Work was also commenced on alteration of the toll sergeants' office and adjoining areas of the administration building at the Toll Plaza to relieve a seriously congested condition.

A larger office was designed and new inter-communication and central control facilities were authorized that would modernize the complete control center with latest types of electric equipment. It is expected that the entire installation will be completed in October or November of 1957.

## Earthquakes

On March 22, 1957 a series of earthquakes shook the San Francisco Bay area as a result of earth displacement along the San Andreas Fault, with a Richter magnitude measuring the earth's energy of 5.5. This compares with 8.3 for the San Francisco earthquake of 1906.

Only minor vibrations were set up in the bridge span. The maximum double amplitude vertical movement was 5.6 inches. This compares with a similar movement of 130 inches during a wind storm of December 1, 1951, before installation of bottom lateral bracing, designed to eliminate any torsional vibrations.

In the March 1957 quake, there was not the slightest semblance of damage to the bridge structure, the towers, foundations or anchorages. A few windows were broken at the toll plaza and in the administration building.

Mr. Clifford E. Paine, Consulting Engineer, was asked to report on the earthquake situation and, in his reply, stated that the bridge had reacted in accordance with anticipated behavior, since it was designed with a factor of safety, amply above the safe limit for the maximum earthquake force that could be conceived for the area, based upon past experience.

This earthquake served as a further test by nature of the durability and strength of the Golden Gate Bridge. □

SIXTY YEARS OF SERVICE

# Highlights from Annual Reports

YEAR: 1957 — 1958

## Report of the President

It was another year of successful operation in which the bridge, the economic life-line between San Francisco and the great Redwood Empire to the north, continued to emphasize its financial stability as traffic reached an all-time new high mark.

This was accomplished under the stimulation of population growth and a high level of business activity in the areas served by the bridge.

And at the close of the 21st year of operation, it is more apparent than ever that there should be no reason for concern over the ability of the bridge to meet all of its current costs and its future obligations of bond interest and redemption, with the present schedule of tolls, barring the unforeseen that would affect traffic and revenue, and abnormal expenditures.

During the period of more than two decades, the bridge has fulfilled all of the expectations of its eminent board of designers and chief engineer. The test of time has proven their theories, verified the wisdom of their decisions and establish a lasting monument to their pioneering courage.

## Looking Forward

Growth of population in counties to the north of San Francisco has been accelerated by improved highways, and planning for better transportation facilities. Commercial and industrial activities have likewise been increasing and will continue to go forward in the future at a rate commensurate with population.

All of these things point toward the necessity of formulating long range plans for the Golden Gate Bridge that will meet the potential problems well in advance of the time they occur in the future. As growth of population and industry continues to the north, highway traffic volume will increase accordingly. This traffic will have to be handled either by the Golden Gate Bridge, by some other structure, or be permitted to stagnate.

## Report of the General Manager

A total of 16,408,399 vehicles cross over the bridge in the 1957-58 fiscal year, an overall increase of 4.67% over the prior year. Average daily number of vehicles for the year just ended was 44,955.

The above figures illustrated significantly the tremendous traffic increase on the bridge when compared with the first year of operation when 3,311,512 vehicles used the bridge, a daily average of 9,072.

## Revenue

For the year, revenue increased to $4,219,181, an uptrend of 2.3%.

The one millionth commutation ticket was sold on February 9, 1958, to Robert E. Byard of Mill Valley, who has been a commuter for eleven years.

---

*In the rest of the World . . . .*

## HISTORICAL SNAPSHOT

**Events.** *Alaska becomes 49th state. First U.S. satellite launched, weight: 30.8 lbs. U.S. nuclear submarine passes under North Pole ice cap. U.S. space agency, NASA, established. Elvis reports for two years' military service.*

**Births.** *Madonna, Michael Jackson.*

**Books.** *Capote, Breakfast at Tiffany's; Kerouac, Dharma Bums; Pasternak, Dr. Zhivago.*

**Films.** *Gigi, The Hidden Fortress, Mon Oncle, Touch of Evil, Vertigo. Television. New programs include, Sea Hunt, 77 Sunset Strip.*

**Music.** *Connie Francis, Who's Sorry Now; National Academy of Recording Arts and Sciences formed, first "Grammy" awards. First stereo records sold in U.S.*

**Everyday Life.** *The Hula-Hoop is invented by Wham-O. Bank of America launches the first multi-purpose credit card. So does American Express.*

# 1958

## "Letters to the Bridge"
### Special Moments at the GGB

During our research we came across many interesting letters written "To The Golden Gate Bridge" over the years. This was not business correspondence, but letters from individuals about a highly personal experience of which the Bridge was a part.

Sometimes it involved driving or walking on the span itself. Often it was from one of the "inspiration points" (there are many)— under the great orange bridge from Ft. Point, or aboard a tour boat, even overlooking it from the Marin Headlands. There's something about the Bridge that inspires special moments in our lives.

Maybe you've had such an experience — a golden moment on the Golden Gate Bridge. Maybe you took a prize-winning photo on one of those remarkable days when light and weather combined in a special way. Perhaps you became engaged at mid-span, or celebrated a birthday or anniversary with a memorable walk or run or bicycle ride over and back.

Maybe you just stood nearby on a brilliant day and were inspired to change your life, start a business, repair a relationship, write a poem, change the world.

You get the idea.

If that happened, we'd like to hear about it because we intend to include some of these real-life experiences in our next edition. We believe this suits our philosophy of "let the people tell the story." We'll be adding your story to the glorious history of the world's most beautiful bridge. We believe one enriching human experience compliments another.

If your story is chosen, we'll send you a copy of the new edition that includes your "Letter to the Bridge." Please limit your story to 200 words or less. (That's quite a bit, really.) We reserve the right to edit all submissions. The address:

"Bridge Letters"
72 Locust Avenue
Mill Valley, CA 94941-2131
USA

## Geologist Report

At the request of the Board, a report was submitted by a group of eminent scientists on geological status of the earth structure in the vicinity of the bridge towers and anchorages. The group was comprised of Professors Francis J. Turner and Perry Byerly of the University of California, and Rear Admiral William M. Gibson, U.S. Coast and Geodetic Survey, retired.

In their separate report, Professors Turner and Byerly emphasized: "We re-endorse and confirm our report of 20 June 1955 in every detail, without revision, modification or qualification of any kind."

In that prior report to which they referred, it was found: "In accordance with past history, we assume that during the next century the San Francisco Bay region will repeatedly be affected by earthquakes, and that a few of these will be severe. We believe, however, that well-designed and properly built structures found on coherent bed-rock, such as the Golden Gate Bridge, will withstand these shocks with minimum damage. We conjecture that the bridge, had it existed in 1906, would have survived the earthquake of that year without substantial damage."

## Structural Improvements

During the year, several structural improvement projects were completed and others initiated on the bridge structure and at the toll plaza.

Repairs to Pylon S-1 stringer bearings were completed on January 17, 1958.

A new intercommunication system was completed in the toll lanes and the Toll Sergeant's office, giving a modern and efficient facility with more space and convenience in the central control office.

## Bridge Appraisal

An appraisal of the bridge was made by the District's engineer, and as of January 1, 1958, it was estimated that actual construction cost of the structure from abutment to abutment would be $111,373,000. The appraisal value, including financing, engineering and administration, less accrued depreciation, was set at $117,189,000.

These figures compare with original cost of $24,471,237 plus subsequent additions and Betterments of $3,703,403, or a total of $28,174,640 without engineering and financial costs.

## Approach Road Improvements

The final link in a divided freeway from the Golden Gate Bridge to Santa Rosa was opened on September 25, 1957, making it possible to drive that 50-mile distance with only four traffic control signals, at Greenbrae, Hamilton Field, Ignacio and Novato in Marin County. □

# SIXTY YEARS OF SERVICE
# Highlights from Annual Reports
## YEAR:1958 — 1959 and 1959 — 1960

*[Editor's Note: For two years, this one and the next, the Annual Reports contained only the Report of the Auditor, which we have not been including in these Highlights, and the financial reports of the District. There was no Report of the President nor Report of the General Manager. In each of these years, traffic on the Bridge increased by more than a million vehicles.]*

## Traffic and Revenue

The total vehicles crossing the Golden Gate Bridge for the fiscal year ending June 30, 1959 was 17,592,396, an average of 48,198 per day.

Total operating revenue for the period was $4,509,698. ☐

---

*In the rest of the World . . . .*

### HISTORICAL SNAPSHOT

**Events.** *Hawaii becomes 50th state. First seven astronauts picked for U.S. space program. Nixon visits USSR, engages in "kitchen debate" with Khrushchev. Later, Khrushchev visits U.S. is refused admission to Disneyland "for security reasons."*

**Music.** *Polenc, La voix humaine (opera); Gypsy, Sound of Music open on Broadway. Berry Gordy founds Motown Records.*

**Deaths.** *Cecil B. de Mille, Raymond Chandler, Frank Lloyd Wright.*

**Books.** *Gunter Grass, The Tin Drum; Alan Stilltoe, The Loneliness of the Long Distance Runner. Sports. Jack Nicklaus wins U.S. Amateur Championship at age 19.*

**Films.** *Ben Hur; The Four Hundred Blows; Hiroshima, mon Amour; Some Like it Hot.*

**Everyday Life.** *Invention of pantyhose. Sony of Japan introduced transistor television set.*

**1959**

---

# The Golden Gate Bridge "Experience"

*If you've visited the Bridge in person or just in these pages, you know it has many personalities. Whether in days of sunlit vistas or brooding fog, this structure reveals as many moods as there are hours.*

*While the "mini climates" of the Bay Area provide fascinating and often contradictory backdrops, it is the Bridge herself which seems to change before our very eyes. Does this phenomenon occur by design? Perhaps. Here are some facts about the sights and sounds of Everybody's Favorite Bridge.*

Acclaimed as one of the world's most beautiful bridges, there are many different elements to the Golden Gate which make it unique. With its tremendous towers, sweeping cables and great span, the Bridge is a sensory beauty featuring color, sound and light.

## ART DECO THEME

The original plans submitted by Chief Engineer Joseph B. Strauss called for a hybrid cantilever and suspension structure across the Golden Gate. This plan was generally regarded as unsightly and a far cry from the elegant, understated lines that define the Bridge today. After Strauss submitted his first design, Consulting Engineer Leon S. Moisseiff theorized that a long span suspension bridge could cross the Gate. A suspension structure of this length had never been tried before.

Even after Moisseiff and Strauss began to refine the new design, it was until Consulting Architects Irving F. Morrow and his wife Gertrude C. Morrow joined that project that the art deco styling began to take shape. The Morrows added the consistent, yet subtle art deco elements which now embody the Bridge. They simplified the pedestrian railings to modest, uniform posts placed far enough apart to allow motorists an unobstructed view. The lightposts took on a lean, angled form. Wide, vertical ribbing was added on the horizontal tower bracing to accent the sun's light on the structure. The rectangular tower portals themselves decreased on ascent, further emphasizing the tower height. These architectural enhancements define the Golden Gate Bridge's art deco form. It is this form which is known and admired the world over.

## LIGHTING THE BRIDGE

Consulting Architect Irving F. Morrow wrote a "Report on Color and Lighting" to Chief Engineer Joseph B. Strauss on April 6, 1935 indicating that the two most important factors in lighting the Golden Gate Bridge are: 1) The enormous size of the project; and 2) The tremendous scale and dignity of the project. Morrow carefully weighed these considerations as he designed his lighting scheme, one which would even further accent the uniqueness of the Golden Gate Bridge.

Because of the Bridge's great size, Morrow did not want the same intensity of light on all of its parts. The effect would seem too artificial. The towers, for example, were to have less light at the top so they would seem to soar beyond the range of illumination. Further, because of the sale and dignity of the Bridge, Morrow believed tricky, flashy or spectacular lighting would be unworthy of the structure's magnificence. Thus, he selected low pressure sodium vapor lamps with a subtle amber glow for the roadway, providing warm, non-glare lighting for passing motorists. The lamps were the most modern available in 1937. Forty-five years later, the original low pressure sodium roadway lights were replaced with high-pressure sodium vapor lamps. These modern lamps provide improved lighting at a lower cost. To preserve the original warm glow, the new lampheads have a plastic amber lens. One of the original lamps is still burning at the Bridge behind the Roundhouse Gift Center just east of the Toll Plaza.

The tower lighting, as originally envisioned by Morrow, was not installed during the construction of the Bridge due to budgetary constraints. However, in 1987, as the climax to the 50th Anniversary evening fireworks ceremonies, the Bridge towers came to life with light. Just as Morrow had envisioned, the new lighting made the towers seem to disappear into the evening darkness, further accenting their great height.

## INTERNATIONAL ORANGE

The Golden Gate Bridge has always been painted orange vermillion, deemed "International Orange." Rejecting carbon black and steel gray, Morrow selected the color because it blends well with the span's natural setting. If the U.S. Navy had its way, the Bridge might have been painted black and yellow stripes to assure greater visibility for passing ships.

Painting the Bridge is an ongoing task and the primary maintenance job. The Bridge paint protects it from the high salt content in the air which rusts and corrodes the steel components. Many misconceptions exist about how often the Bridge is painted. Some say once every seven years, others say from end to end each year. Actually, the Bridge was painted when it was originally built with a red lead primer and a lead-based top coat. For the next 27 years, only touch up was required. By 1965, advancing corrosion sparked a program to remove the original paint and replace it with an inorganic zinc silicate primer and acrylic emulsion topcoat.

## FOG HORNS

The original Bridge fog horns, one at mid-span and one at the south pier, remained in use for nearly 50 years. Their deep, baritone sounds guided hundreds of thousands of vessels safely through the Gate, and forewarned San Franciscans when fog was rolling in to envelope The City.

Unfortunately, in the late 1970's, the two-tone fog horn at mid-span stopped working. One of the horn's two air valves gave way and the two-tone horn became a one-tone horn. But since the mechanism was so old, replacement parts were impossible to find. The hobbled horn continued to sound its one-tone beacon until 1985, when both of the original horns showed signs of wear, tear, and exposure to the elements. The original fog horns were replaced with new horns manufactured by the Leslie Air Horn Company. The new horns, while differing in frequency or tone from one another, are all single-toned horns which operate, like the originals, with compressed air.

The fog horns operate, on average, two and a half hours a day. During March, you'll hear them for less than half an hour a day. But during the Bay Area's foggy season, July through October, they sound over five hours a day.

Small vessels that do not have radar still use the Bridge fog horns as a guide when visibility in the Golden Gate Strait is low. Each horn has a different pitch and marine navigational charts give the frequency, or signature, of each fog horn. Vessels heading into the Bay steer left of the south pier horn and right of the mid-span horn. Outbound vessels stay to the right of the mid-span horn.

## BEACONS

The Bridge is also equipped with navigational and warning lights for travelers by sea and air. Originally, a red rotating aircraft beacon showed on the top of each tower. In 1980, they were replaced with 360 degree flashing red beacons. The Bridge main cables are also marked with red cable outline lights. For seafaring vessels, there are red navigation lights on the south pier fender and white and green lights below the deck at mid-span. □

## SIXTY YEARS OF SERVICE

# Highlights from Annual Reports

## YEAR: 1959 — 1960

*[Editor's Note: For two years, this one and the year prior, the Annual Reports contained only the Report of the Auditor, which we have not been including in these Highlights, and the financial reports of the District. There was no Report of the President nor Report of the General Manager. In each of these years, traffic on the Bridge increased by more than a million vehicles.]*

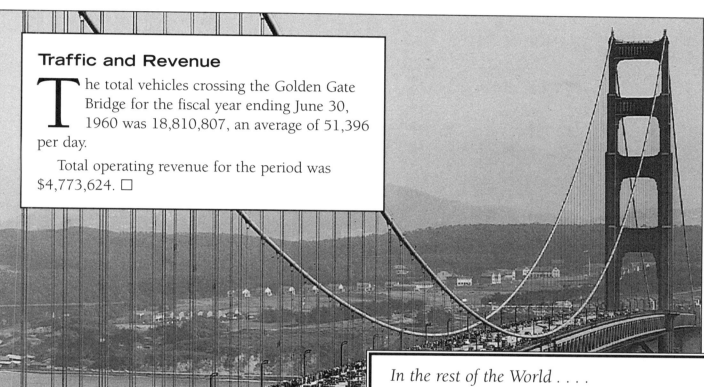

### Traffic and Revenue

The total vehicles crossing the Golden Gate Bridge for the fiscal year ending June 30, 1960 was 18,810,807, an average of 51,396 per day.

Total operating revenue for the period was $4,773,624. □

*In the rest of the World . . . .*

## HISTORICAL SNAPSHOT

*The Bridge was conceived and built during an important, expanding time in history. It's birth was difficult; it exists in a real world. These are a few of the events, near and far, that happened concurrently with its development and growth. As you'll see, from time to time events occuring far from San Francisco Bay held influence over the business of the Golden Gate Bridge.*

**Events.** *U.S. U-2 spy plane flown by Francis Gary Powers, downed over USSR. Paris summit meeting canceled because of the incident. Sen. John F. Kennedy elected President, narrowly defeating Richard Nixon. First U.S. weather satellite, Tiros I, launched. First oral contraceptive, "The Pill," approved by FDA. California passes New York as most populous state. British surgeons develop heart pacemaker.*

**Books.** *John Barth, The Sot-Weed Factor; John Updyke, Rabbit Run.*

**Drama.** *Robert Bolt, A Man for All Seasons; Eugene Ionesco, The Rhinoceros.*

**Architecture.** *Brasilia, new capital of Brazil is opened. Frank Lloyd Wright's Guggenheim Museum opens in New York.*

**Music.** *Camelot, Oliver! open on Broadway. Johnny Mathis, The Shadow of Your Smile.*

*Edit Piaf, Non, je ne regrette rien. First performance by John Lennon, Paul McCartney, George Harrison, Pete Best, as The Beatles, in Hamburg, Germany.*

**Films.** *The Apartment, La Dolce Vita, The Magnificent Seven, Never on Sunday, Psycho.*

**Everyday Life.** *The Twist is latest dance craze. Go-Karts become popular. Pentel felt-tip pens go on sale in Japan. Haagen-Dazs ice cream created in the Bronx, New York.*

**The Golden Gate Bridge** *was trying to find ways to increase capacity and studying something called BART.*

## 1960

# TIPS FOR VISITORS

## Information You Should Know

The Bridge can be a very busy place, particularly during the summer months. It is estimated that over nine million people from around the world visit the Golden Gate Bridge each year.

## Weather

The climate is temperate marine and generally mild year-round. Daytime temperatures range from 40 degrees (F) in the winter to 75 degrees (F) in the summer. Morning and evening fog rolls in during the summer months, but generally burns off by midday.

## Vistas & Parking

There are two main vista points located at the Bridge. One on the northeast side and the other in on the southeast side. Parking is available at both sites, including parking spaces for persons with disabilities. Our neighbor to the north and south, the Golden Gate National Recreation Area, offers additional viewing areas.

## What Not to Miss

The following attractions are located on the southeast side of the Golden Gate Bridge Toll Plaza and may be reached most easily by parking in the East Parking Lot. To access the East Parking Lot: Traveling northbound, take the last San Francisco northbound exit off Highway 101. Traveling southbound, proceed through the far right toll lane (west side). Take an immediate right exiting Highway 101. Make another immediate right onto a roadway that passes underneath the toll plaza and directly into the East Parking Lot.

### Joseph B. Strauss Statue

**The Joseph B. Strauss statue is a constant reminder to all of the determination it took to build this great structure. The statue represents the District's continued recoginition of the man who refused to give up until his vision became a reality.**

## Main Cable Cross-Section

Adjacent to the Strauss statue is a section of the main cable of the Golden Gate Bridge. The cross-section demonstrates the magnitude of this incredible engineering feat. Various construction statistics are also displayed.

## Glorious Golden Gate Gardens

Since the 1960s, the renovated garden areas have been the backdrop showcasing the historic Bridge to visitors from around the world. As visitors step back from the Strauss statue, they are faced with yet another visual treat: the immaculate gardens. On less then five acres, the annual and perennial flower beds and manicured hedges accent the brick sidewalks inviting guests to investigate or wander up or down a path to view the Bridge from a different perspective.

## Roundhouse Gift Center

The District's Gift Center is located in the historic "RoundHouse" building. Constructed in 1938, the Roundhouse was originally a restaurant for passing motorists. In 1973, it became District offices. Then in 1987, the building was transformed into the Gift Center. Open seven days a week between 9 a.m. and 5 p.m.(8:30 a.m. to 7:30 p.m. during summer months), the Gift Center offers a variety of information and souvenirs of the Golden Gate Bridge.

## The Cafe

In addition to visiting the Roundhouse Gift Center, visitors may enjoy food and beverages from a varied menu at the walk-up Cafe, adjacent to the Strauss statue.

## Pedestrian Sidewalk Hours

Visitors to the Golden Gate Bridge may use the east sidewalk, near the Strauss statue to walk to the north side of the span. The sidewalk is open from 5:00 a.m. to 9:00 p.m. seven days a week.

## Bicycle Access

Bicycles are permitted to cross the Golden Gate Bridge as follows:

| **Weekdays** | **Weekends & Holidays** |
| --- | --- |
| 5:00 a.m. to 3:30 p.m. via the east sidewalk | 5:00 a.m. to 9:00 p.m. via the west sidewalk |
| 3:30 p.m. to 9:00 p.m. via the west sidewalk | 9:00 p.m. to 5:00 a.m. via the east sidewalk |
| 9:00 p.m. to 5:00 a.m. via the east sidewalk | |

SIXTY YEARS OF SERVICE

# Highlights from Annual Reports

## YEAR: 1960 — 1961

### Report of the President

The Bridge has always been operated, since its opening day in 1937, without cost to the taxpayers and it does not receive subsidies of any kind from the State, Federal Government, or any other source.

A long-range plan to increase the vehicle-carrying capacity of the Bridge by 25 per cent, without cost to the taxpayer, was approved and implemented.

A proposal to place rapid transit trains on the Bridge was received from the San Francisco Bay Area Rapid Transit District, and an investigation as to the engineering feasibility was first approved.

### Traffic

For the first time, the annual Bridge traffic exceeded 20,000,000 vehicles. It is significant that this represents almost a 100 per cent increase or double the Bridge traffic in the past ten years. The 1950-51 traffic volume was 10,153,493.

### Legislation

The Legislature adjourned its 1961 regular session on June 16. Major bills affecting the Bridge District — SB 853 and AB 2739, creating a Golden Gate Transportation Commission by merging the Golden Gate Bridge and State toll bridges in the San Francisco Bay Area — were defeated in the Senate Transportation Committee. The bills represented a modified version of a somewhat similar measure in the 1959 Legislature to create a Golden Gate Authority of bridges, airports and port facilities, which was also defeated in the Legislature.

### Future Planning

Directors reviewed and recommended a course of action, composed of three phases, that would enlarge the capacity of the Bridge to carry vehicles.

Chiefly, these had to do with a proposed widening of the Marin viaduct, or north approach structure; traffic control on the span itself by reversing the direction of a lane at commute and other peak hours; and a realignment and enlargement of the toll collection facilities at the south end of the Bridge.

### Revenue

Revenue for 1960-61 reached the record annual total of $4,871,196, a gain over 1959-60 of 2.4 per cent.

The revenue increase was accomplished in the face of two toll reductions. The first was a reduction from $9 to $8 in the 52-trip commute ticket book, effective September 1, 1960. The second was a reduction in the rate for two-axle trucks with single rear wheels (including light delivery trucks) from 50 cents to 25 cents, effective May 1, 1961.

---

*In the rest of the World . . . .*

## HISTORICAL SNAPSHOT

**Events.** *John Kennedy becomes 35th U.S. President, later accepts full responsibility for "Bay of Pigs" fiasco. Government sources advise U.S. families to build or buy bomb shelters. First "skyjacking" occurs. First American in space, sub-orbital flight by Cdr. Alan Shephard.*

**Books.** *Jos. Heller, Catch 22; J D Salinger, Franny and Zooey.*

**Music.** *Mancini, Moon River. Bob Dylan debuts. Country Music Hall of Fame founded.*

**Films.** *Breakfast at Tiffany's, Jules et Jim, 101 Dalmatians, A Taste of Honey, Yojimbo.*

**Television.** *The Defenders, a top program.*

**Sports.** *Roger Maris breaks Babe Ruth's record with 61 home runs.*

**Everyday Life.** *After 78 years, the Orient Express (Paris- Bucharest) made its last run.*

# 1961

### Rapid Transit

The Board received from the San Francisco Bay Area Rapid Transit District a copy of a report prepared by its traffic engineer, that proposed addition to facilities "to accommodate two tracks of conventionally supported rapid transit on the suspension spans of the Golden Gate Bridge."

Engineer Clifford E. Paine, chief designer of the Bridge, was authorized to review the report and give his findings to the Board.

### Looking Forward

As the Bridge District enters its Twenty-fifth year of operation it looks forward with pleasure, confidence and optimism to the Silver Jubilee on May 27-28, of the "Bridge that couldn't be built."

### Report of the General Manager

The Golden Gate Bridge has the lowest average toll per vehicle of any toll bridge in the San Francisco Bay region, on the basis of 1961 traffic and revenue figures. Its average toll is 30 percent below its nearest toll competitor, the San Francisco-Oakland Bay Bridge.

Three all-time-high traffic marks were recorded during the fiscal year. All records were set on Sundays. These included July 24, 1960, 73,771 vehicles; August 2, 1960, 74,991 vehicles; and June 25, 1961, 75,487 vehicles.

The lowest traffic ever recorded on the bridge occurred on December 8, 1937, in the opening year, when only 3,878 vehicles crossed the span.

### Maintenance

During the year, work began on the suspender cable bolt replacements, which tied the 500 suspenders of the span to the main cables. The work does not in any way imperil the safety or stability of the bridge and is being carried out under the strict observation, engineering inspection and control of safety engineers.

Replacements are being made by the District's own skilled and experienced maintenance forces, which results in very material savings in costs.

A new 500-gallon fire truck was delivered to the District on August 15th and tests have proved satisfactory in every respect.

### Free Tow Truck Service

Bridge service crews responded to 4,383 calls from motorists disabled on the bridge or its approaches during the year.

Under the District's policy, emergency vehicles carry gasoline and water to stranded motorists and provide free towing service. Vehicles out of fuel are supplied with three gallons of gasoline at cost.

Of those helped by the program, 1,035 cars were supplied with gasoline, 974 were towed, 775 received tire changes, 240 were serviced for water, and 1,359 stalled by battery or motor trouble were started. □

## SIXTY YEARS OF SERVICE

# Highlights from Annual Reports

## YEAR: 1961— 1962

### Report of the President

The 25th year of operation of the Golden Gate Bridge closed on June 30, 1962, with a phenomenal record for its first quarter of a century in public service.

During that period, except for wartime years, annual traffic continued on its upward trend from 3,326,521 for the first year of service to double that volume in the ninth year; triple during the fourteenth year; quadruple in the 18th year; and close to sevenfold during this 25th year, reaching a total of 21,377,944 vehicles.

### In Retrospect

This great bridge has been subjected to adversity since its inception, but now stands firmly as a world-renowned monument to its financial sponsors, its engineering designers and the people of the counties in its District whose foresight, faith and confidence made its creation possible.

Early critics who said it could not be built, and those who said it would be a financial failure if built, have long since been disproven. Their ill-conceived prophesies have been discredited by the testing laboratory of time.

Engineering and scientific inspection made at appropriate intervals show the structure to be sound and its foundations solid. The generous factor of safety that was provided by its designers has not proven to be wasteful.

Not only the steel and concrete performed as anticipated, but the financial structure has out-stripped the most optimistic predictions. All costs have been paid out of earnings, including operation, maintenance, additions and Betterments, and bond redemption and interest.

Let the record speak for itself — not in voice raised against the self-serving interests of a few critics, but directed to the greater cause which subordinates all others — the public welfare of the people of the counties of this District and the great abundance of visitors who seek out the attractions of the Golden Gate Bridge.

In this light the Board of Directors, the managerial staff, and the loyal employees of the District can look upon their accomplishments with much pride and personal satisfaction for a job well done.

---

*In the rest of the World . . . .*

## HISTORICAL SNAPSHOT

*[It is noteworthy that the Annual Report marking the first quarter-century of service by the Golden Gate Bridge would pause for a brief backward glance at its accomplishments, at the challenges it met and conquered, at its place in the history of the region, and its courageous consideration of the future it boldly faced. There's more than a little pride revealed here — pride in work well done, responsibilities fulfilled, obligations met all the while looking forward, preserving for unknown tomorrows the function, art and soul of the world's most beautiful bridge.]*

**Events.** *In April, the Cuban missile crisis resulted in a naval blocade. President Kennedy lifted the blocade in November. Lt. Colonel John Glen became first American to orbit earth. Hugh Hefner published a book, The Playboy Philosophy. Sedative thalidomide banned.*

**Music.** *Shostakovich, Symphony No. 13. Tony Bennet, I Left My Heart in San Francisco. Elvis, Return to Sender.*

**Books.** *Ken Kesey, One Flew Over the Cuckoo's Nest; Katherine Anne Porter, Ship of Fools.*

**Films.** *Lawrence of Arabia; Dr. No, first Connery Bond; To Kill a Mockingbird.*

**Deaths.** *William Faulkner, Marilyn, Eleanor Roosevelt.*

**Everyday Life.** *Johnny Carson takes over The Tonight Show. First ring-pull drink cans on market.*

# 1962

## Rapid Transit

During the past year there have been important events bearing upon the future welfare of the San Francisco metropolitan area and the counties of the Golden Gate Bridge and Highway District. Final planning studies of the San Francisco Bay Area Rapid Transit District were concluded in preparation for presentation to the voters at an election in November 1962.

Provision had been made in the original program for a rail rapid transit line between San Francisco and Marin County, using the Golden Gate Bridge as a means of crossing the Bay. The consulting engineer for the Transit District had declared such use of the bridge to be feasible, but the Bridge District's engineer, Mr. Clifford E. Paine, took exception and recommended against installation of the rail system on the bridge.

In the interest of public welfare and safety, the conflicting findings of these two engineers were submitted to an Engineering Board of Review for final decision.

It was the finding of this Board after review of prior reports and independent analysis, that it would be inadvisable to install rapid transit tracks on the Golden Gate Bridge. This decision was accepted by both the Bridge District and the Transit District as final. As a consequence, rail rapid transit to Marin County was excluded from the projected regional system.

## Planning

During the year, several phases of the long range planning program were processed in an effort to meet the increasing demand of traffic on the bridge.

Widening of Marin County end of the Golden Gate Bridge progressed satisfactorily toward completion; steps were taken to establish reversal of center lanes on the bridge to provide four lanes in the direction of peak hour flow; engineering soil tests were made in the Toll Plaza area in contemplation of modernization; several plans of Toll Plaza design were considered; consideration of planning for a second Marin County bridge was placed in committee; and studies were initiated to determine the feasibility of adding more vehicular traffic lanes to the present bridge.

## Report of the General Manager

With the end of the 1961-62 fiscal year, the Golden Gate Bridge has completed 25 years of service with a record of achievement and financial success that exceeded all predictions of its sponsors.

Not only has it established new records of traffic year after year, but it has paid its own way from the very first opening day in May 1937, without costing the taxpayers a penny.

This great bridge — still after a quarter of a century the world's greatest suspension bridge with the longest single suspended span, 4,200 feet — has magnificently stood the test of time. It has stood through the storms of adversity in initial planning and financing, the tremendous forces of nature through high waves and currents, and economic depression and world conflict, one of which has left a lasting scar.

These are momentous achievement that should be looked upon with serious contemplation in considering the probability of future durability of the bridge.

## Traffic and Revenue

Total vehicles (for the year), 21,377,944 for a daily average of 58,570. Total revenue: $4,820,046. Average toll per vehicle: 0.2246.

## Maintenance

Work on replacement of clamps on the suspender cables on the bridge structure proceeded during the year toward an ultimate of 516 clamp assemblies of new welded design and superior protection against corrosion.

Repairs to the north and south tower maintenance elevators were completed during the year by the Otis Elevator Company.

In all other respects the normal bridge preventative maintenance program of painting and repair proceeded on schedule.

## Toll Plaza Modernization

As part of the long range planning program, studies were conducted to explore several methods and designs of modernizing the Toll Plaza so as to accommodate the rapidly increasing vehicular traffic volume.

Among the plans considered was a proposal to relocate the toll gates to the west and to the south, and to replace the present Administration building with a new structure to be located at another site at the plaza area.

A second plan, for a split Toll Plaza provided for preserving the Administration Building, constructing a new set of 12 toll gates to the west thereof for southbound traffic and then replacing the present toll gates with new facilities for northbound traffic. The Administration building would remain in its present location and would be between the two sets of toll gates.

The latter plan was approved in principle, but final action depends upon results of further study.

## Marin Approach Widening

During the past year, one of the important long range planning projects was initiated, whereby the north end of the bridge would be widened on the west side by 20 feet at the abutment, tapering down to zero near the cable anchorage housing, thus giving a roadway width of 80 feet at the bridge entrance, gradually reducing to the 60 foot width on the suspended portion of the bridge.

This project was designed to reduce a serious accident hazard. In June 1962 the project was 75% completed.

## Automatic Toll Collectors

Four automatic toll collecting machines in use since February 1960, were retained at the toll gates on a month-to-month lease basis for continued evaluation. All of the machines were replaced by the manufacturer with improved models during the latter part of the year.

Although these machines cannot be used to maximum potential during the peak hours due to the high ratio of commute ticket use, they serve a useful purpose during off peak hours for cash toll collection and substitution for manual collectors during short relief periods.

Ultimate disposition of the machines will depend upon final design of the toll gate facilities and type of commutation toll rate medium.

## Second Marin County Bridge

During the year a preliminary report was completed by the State Division of San Francisco Bay Toll Crossings on feasibility of construction of a second Highway crossing between San Francisco and Marin County.

The findings were that a high level suspension bridge would be the only feasible manner to make the crossing.

It was proposed that the San Francisco terminal would be located between piers 37 and 39, opposite Battery and Sansome Streets and would be routed via Angel Island, thence to the Tiburon Peninsula on the north.

The bridge would be five miles in length and nine miles overall, including approaches. It would be designed for two decks with provision for a double track rapid transit line on the lower level together with two vehicular traffic lanes. The upper deck would be for vehicular traffic.

Because of the public opposition to the preliminary plan, the Governor [E. G. "Pat" Brown] brought the studies to a halt and asked the State Highway Transportation Agency to prepare a prospectus for a San Francisco Bay Area-wide study of surface transportation, including an overall plan for highways, rapid transit, bridges and under-water tubes. □

# CELEBRATION II TWENTY-FIVE YEARS

*[The 25th anniversary of the Bridge opening was modest considering the early political, social and engineering achievement of the Gate span. As these notes from the Annual Report for the year reveal, there was a luncheon, some old cars and a plaque. It probably seemed adequate at the time, but as history will show, supporters of the Golden Gate Bridge would revel in their support, admiration, and love twenty-five years later, when hundreds of thousands of people would assemble to share in a triumphant 50th Anniversary Celebration.]*

Observance of the 25th anniversary of the Golden Gate Bridge began with a luncheon at the St. Francis Hotel on May 25, sponsored by the Redwood Empire Association. More than 400 persons attended, including present and past directors of the Bridge District, and many public officials of the counties of the District.

Mr. Jack Craemer, President of the Redwood Empire Association, opened the luncheon and Director Dan E. London of the Bridge District acted as toastmaster. An inspiring speech was delivered by Dr. Robert Gordon Sproul, President Emeritus of the University of California.

On May 27th, there was an antique car parade which proceeded from Vista Point in Marin County, across the bridge and into the Presidio on the San Francisco end.

A bronze commemorating plaque was received on behalf of the Bridge District.

The first quarter century of bridge operation was brought to a close with a highly successful conclusion, laying the groundwork for beginning the next step toward the Golden Anniversary of the Golden Gate Bridge, by which time the financial obligations of the District will have been fully discharged and traffic saturation will have long been reached. □

## SIXTY YEARS OF SERVICE

# Highlights from Annual Reports

## YEAR: 1962 — 1963

### Report of the President

The year saw the completion of the first step in a long range planning program designed to increase the vehicular capacity of the bridge. This was the widening of the Marin approach structure at the north end of the bridge to permit a smoother transitional flow of traffic between the freeway on the north and the span.

The project was immediately followed by an experimental traffic control system of providing four of the six lanes across the entire bridge structure in peak hours for the heaviest flow of traffic. This plan, while temporary and accomplished by manually reversing the center lane with traffic dividers, met with enthusiastic approval.

The system was subsequently extended from the Toll Plaza south along the San Francisco approach through the cooperation of the California State Division of Highways, which has jurisdiction over the approach.

In the meantime studies have been pursued for a permanent overhead traffic control system, which the board of directors has already approved in principle.

The board of directors has initiated action looking towards a possible engineering feasibility investigation of a lower deck on the Golden Gate Bridge for light vehicles. It has also urged continued studies, originally begun and then discontinued by the state of California, for a second crossing between San Francisco and Marin County, which is considered inevitable. Likewise, it has advocated immediate action by the state in widening the main San Francisco approach to the Bridge.

All recent surveys on the future development of the San Francisco bay area have indicated that the north bay counties, served by the Golden Gate Bridge, will continue to have rapid growth in population. Therefore, immediately planning, which is obviously the only intelligent approach to formulation of a program to meet the future transportation needs of this vast territory, cannot be delayed.

### Report of the General Manager

With California emerging as the No. 1 state in population in the United States and continuing to boom, the growth and development of the state and particularly the region northwest of San Francisco Bay was reflected in Golden Gate Bridge statistics for the fiscal year ending June 30, 1963.

Bridge traffic increased over 1961-62, reaching a total of 22,514,356 vehicles, a gain of 5.32%. Total revenue amounted to $5,040,421. During the past ten years, there was a growth average in excess of 1,000,000 a year.

For the purpose of estimating facility requirements and traffic growth, a maximum capacity per bridge traffic lane of 1,500 vehicles has been adopted during the past years. That figure was exceeded in 1962 both northbound and southbound during the peak hour, with three traffic lanes in each direction on the bridge.

---

*In the rest of the World . . . .*

## HISTORICAL SNAPSHOT

**Events.** *President John Kennedy assassinated in Dallas. Freedom March on Washington D.C., Dr. Martin Luther King's "I have a dream..." speech.*

**Books.** *Betty Friedan, The Feminine Mystique; James Baldwin, The Fire Next Time.*

**Music.** *Pavorotti makes debut at Covent Garden. Beatles, Please Please Me, I Want to Hold Your Hand.*

**Films.** *Cleopatra, The Great Escape, Le Mepris, Tom Jones.*

**Sports.** *Stan Musial retires. Palmer first golfer to win $100,000 in single season. Don Schollander of Santa Clara, Calif first person to break 2-minute mark for 200-meter freestyle swim.*

**Everyday Life.** *Weight Watchers founded in New York. U.S. Post Office introduces Zone Improvement Plan, ZIP codes.*

## 1963

By devoting four traffic lanes to peak hour traffic, the 1963 volume per lane during peak hours was reduced to 1,356 southbound in the morning and 1,321 northbound in the evening.

It is obvious that if the upward trend continues, a maximum capacity of the four lanes will be reached within a short period of time.

## Improvements

*[The General Manager echoed and expanded the President's comments regarding the four-lane traffic control system, including extending to southward along the San Francisco approach.]*

San Francisco city officials cooperated in speeding bridge traffic by lengthening the time for traffic to move along access streets to the bridge approaches.

As another step towards expediting traffic through the toll lanes, the District converted all off-drive side toll collection booths to on-driver side collection. The new booths were designed, fabricated and installed by the District's maintenance force and are temporary, pending planned modernization of the Toll Plaza.

The District had experimented with four automatic toll collection machines over a period of two years but discontinued their use because of the limited value.

Meanwhile other studies were authorized by the board of directors as to the feasibility of a one-way toll collection system on the bridge and of increasing the use of buses on the span as a means of relieving private car congestion.

## Trans-Bay Traffic Study

The State Division of San Francisco Bay Toll Crossings made public during the year a report entitled "Trans-bay Traffic Study" dated November 1962, in which were presented the results of an origin and destination traffic survey conducted in the fall of 1961 on the Golden Gate Bridge and the five State-owned toll bridges in the San Francisco Bay Area.

The report estimates that the Golden Gate Bridge will "exceed its capacity of 25,000,000 vehicles in the year 1967." The District's own estimate is that the Bridge will reach *[that figure]* before the end of 1965.

The report emphasizes that the survey "has rather graphically brought out the fact that the Bay Area can no longer delay in arriving at some decision regarding its trans-bay traffic problems, particularly with regard to the construction of additional bridges across San Francisco Bay."

## Personnel

A major factor in the successful operation of the Golden Gate Bridge down through the years is the loyal and active cooperation of its employees.

The District is proud of the excellent relations existing with the employees and the management acknowledges with sincere gratitude the contributions which they have made through their dedicated service to the bridge and the motoring public.

*[The Annual Reports of the District consistently praised the excellence of its work force, though we have not included each of those commendations in these Highlights.*

*The 1962-63 Report also included a kind of "afterword" which began by praising "The Most Beautiful Bridge in the World" and went on to summarize its stormy early beginnings, its encounters with free-use government vehicles, a major windstorm and its resultant retrofit, a steadfast fiduciary responsibility, and a look at what appeared to be the most critical issue of all — how to respond to the ever-increasing traffic problem which threatened Early Gridlock. In doing so, the Directors and Management of the District revealed a responsive, forward-thinking and sympathetic attitude about the transportation problems which each year were descending with greater force on Californians. — Editors.]* ☐

SIXTY YEARS OF SERVICE

# Highlights from Annual Reports

YEAR: 1963 — 1964

[In a step toward modernization, the Report this year dropped the old headings, "Vehicles," "Revenue," "Maintenance," etc, in favor of a less-formal narrative style. This year also reflects new efforts at art direction, with a "posterized" cover printed on colored stock and the use of more creative design inside.]

## Report of the President

Together with the distinction of having the largest population of all the states, there has come to California the difficult problem of keeping pace with the transportation needs of the expanded population.

To meet present and future demands, a regional master transit system is presently in progress that will provide rapid interurban rail lines between the East Bay cities and San Francisco. To the north, however, there will be no such source of relief, as those counties have been excluded from the rail system due to their inaccessibility by use of existing facilities. Under these circumstances, the Golden Gate Bridge remains as the only major link between San Francisco and the area to the north.

In an effort to cope with continued steady growth, which will soon exceed the rated capacity of the Bridge, much attention has been focused upon means of increasing the traffic-carrying capacity of the Bridge structure itself and its approach facilities. The ultimate solution, however, must be a new bridge.

## Report of the General Manager

A total of 24,023,128 vehicles was recorded for the year. Maximum practical capacity of the Bridge has been estimated at 25,000,000 vehicles. On the basis of the present trend, this figure will be reached by mid 1965. Revenue for the Bridge in 1963-64 amounts to $5,358,257.

During the past ten years, there has been an average annual increase of more than 1,175,000 vehicles handled.

During the year, an extensive program of bridge sidewalk repairs was brought near to completion. This project involved cutting away about 20 inches from the outer edges of the sidewalks on both sides of the Bridge to correct a corrosion condition existing between the concrete and the steel supporting structure. Expansion joints in the sidewalks were also replaced.

A number of minor repairs to the Bridge roadway were found to be necessary as the concrete surface began to show its age and the effects of continuous impact from heavy traffic. Ultimately, a complete resurfacing project will be undertaken to provide a new wearing surface.

Design of restroom facilities at Vista Point at the north end of the Golden Gate Bridge was carried to completion by the State Division of Highways. This improvement will complete the landscaping and provisions for tourist facilities at the point, a popular rendezvous for motorists which is under State jurisdiction.

When Alcatraz Island was abandoned as a Federal prison, President Lyndon Johnson appointed a commission to investigate possible uses of the island for other purposes, among which was a suggestion that it be used as an anchorage for a new San Francisco-Marin bridge.

---

*In the rest of the World . . . .*

## HISTORICAL SNAPSHOT

**Events.** *Civil Rights Act of 1964 signed by President Johnson. Major earthquake in Alaska kills 117. Surgeon General's report on "Smoking and Health" confirms link between smoking and lung cancer, heart disease*

**Music.** *Fiddler on the Roof and Hello, Dolly on Broadway. Roy Orbison, Pretty Woman. The Beatles, Hard Day's Night. The Animals, House of the Rising Sun.*

**Films.** *Dr. Strangelove, A Hard Day's Night, Mary Poppins, Zorba the Greek.*

**Sports.** *Cassius Clay beats Sonny Liston. Boston wins sixth-in-a-row NBA championship over San Francisco Warriors, 4 games to 1.*

**Everyday Life.** *Beatles appear on Ed Sullivan Show. Liz Taylor marries Richard Burton. Summer Olympics held in Tokyo.*

# 1964

## The Bridge Division

These trade unions and organizations represent the employees of the Golden Gate Bridge, Highway and Transportation District who work on The Bridge Division under the direction of the Bridge Manager:

- Transport Workers Union of America, AFL-CIO, Local 250 (Bridge Officers)
- Bay Counties District Council of Carpenters, Local 222
- Cement Masons Union, Local 580 of the International Association of Operative Plasters and Cement Masons
- Local 377 of the International Association of Bridge, Structural, Ornamental, Reinforced Iron Workers, Riggers and Machinery Movers
- International Brotherhood of Electrical Workers, Local 6

- Laborers' International Union of North America, AFL CIO, Local 291
- Automotive Machinists, Local 1305 of the International Association of Machinists and Aerospace Works and Machinists (Bridge Mechanics)
- International Union of Operating Engineers, AFL-CIO, Local 3
- International Brotherhood of Painters and Allied Trades, AFL-CIO, Local 4
- United Association of Journeymen & Apprentices of the Plumbing and Pipe Fitting Industry, Local 38

The Commission, after public hearings, recommended that Alcatraz Island be the site of a monument commemorating the 1945 founding of the United Nations in San Francisco and as a possible anchorage for a second bridge to Marin County.

On August 12, 1963, the San Francisco Board of Supervisors voted approval of a cross-town tunnel leading to a depressed freeway, connecting the Bridge with the Central Freeway. This connection would be in lieu of the former plan to connect the Golden Gate Bridge with the San Francisco-Oakland Bridge via the Embarcadero Freeway.

A new proposal was considered by the San Francisco Board of Supervisors that would carry traffic from the present stub- end of the Embarcadero Freeway at Broadway, by tunnel beneath Telegraph and Russian Hills, then through a tube around the yacht harbor, to emerge in the vicinity of Crissy Field in the Presidio.

Work was completed on the first stage in widening the southbound approach to the Toll Plaza, immediately south of the south abutment to provide better access to the toll gates. The area used for widening was originally a gun-emplacement of the Presidio's old Battery Lancaster which once contained three 12-inch breech-loading disappearing guns. The guns were removed in 1918 and the remaining facilities abandoned except for storage of supplies and emplacement of anti-aircraft guns during World War II.

In Marin, widening of the west side of the Waldo approach from the foot of the grade to Waldo Tunnel was carried out by the State to provide an extra lane for southbound traffic proceeding up the grade. To provide the same benefits for northbound traffic, the east side of the roadway from the bridge to a point beyond Waldo Tunnel near Spencer Avenue in Sausalito was widened.

During a special session of the State Legislature in the Spring of 1964, an enabling bill was passed, creating a Marin Transit District, subject to a majority vote of approval by Marin County voters. It provided that the District, if established, could impose a tax of up to 5 cents per $100 assessed valuation of property in the County; also that the District could be integrated with the San Francisco Bay Area Rapid Transit District in the future.

Another bill was passed by the Legislature appropriating funds to study feasibility of placing rapid transit on the San Rafael-Richmond Bridge. □

## SIXTY YEARS OF SERVICE

# Highlights from Annual Reports

## YEAR: 1964— 1965

### Report from the President

The upward trend in traffic on the Golden Gate Bridge continued unabated during the fiscal year ending June 30, 1965. The total number of vehicles crossing the Bridge during the period totaled 25,282,353.

The Board of Directors has, for a number of years, been examining methods to accommodate this increasing traffic, which has now exceeded the maximum vehicular capacity that had previously been estimated at 25 million annually.

For example, consideration of constructing a second deck on the Bridge was actively pursued during the year, relating to a preliminary feasibility engineering report, which recommended a four-lane lower level for light vehicles only.

Also given consideration was a proposal which entailed double-decking the present Bridge and then building a duplicate double deck on a so-called tandem structure consisting of a third leg to each of the two existing towers. This plan was determined not to be feasible.

To accommodate the ever-increasing traffic, a lane reversal program during peak hours, which in recent years, has proved to be efficient in handling commute vehicles, was extended.

However, studies were undertaken in the last year seeking to improve the method of lane reversal by means of automatic "pop-up" tubes or roadway surface lights.

A project to widen the south end of the Bridge, on the westerly side immediately north of the Toll Plaza, was completed early in 1965. The roadway was increased by 17 feet 6 inches, thereby providing better distribution of southbound traffic from the four lanes on the Bridge into ten toll gates used during the morning peak hours.

Each of (several other) studies and actions is designed to improve facilities and to help the existing Bridge accommodate increased traffic. The ultimate solution, though is a second crossing between San Francisco and Marin Counties, which is still years away.

During the 1964-65 fiscal year, there were 8,000,000 commute autos and 156,000 buses using the Bridge. Buses carried 28% of total commute passenger traffic.

It was estimated by the Bridge District Engineer that if a 10% loss of bus passengers should result from an increase in bus fares, then proposed by Western Greyhound Lines, there would be an increase of approximately 468,000 commute autos on the Bridge annually.

---

*In the rest of the World . . . .*

## HISTORICAL SNAPSHOT

**Events.** *First U.S. combat forces sent to Vietnam. Malcolm X assassinated. Watts riots in Los Angeles cause $200-million damage. Northeast power blackout darkens 80,000 square mile area. First "space walk" by astronauts. Head Start Program begins.*

**Architecture**. *Eero Saarinen, Gateway Arch, St. Louis. Kenzo Tange, Roman Catholic Basilica of St. Mary, Tokyo.*

**Music.** *First audio cassettes released. Petula Clark, Downtown. Tom Jones, It's Not Unusual.*

**Films**. *Doctor Zhivago, The Knack, The Sound of Music.*

**Sports.** *Dodger Sandy Kofax pitches fourth no-hitter of career, vs Cubs.*

**Everyday Life.** *Young men sport shoulder-length hair. Britain adopts metric system. Miniskirts become fashionable. Estimated 350-million viewers worldwide watch Winston Churchill funeral on television.*

# 1965

Hoping to prevent this, the Bridge District proposed to the State Public Utilities Commission that if the bus rate increase should be withdrawn as applied to Marin and Sonoma County commuters, the Bridge District would reduce the Bridge bus toll in an amount equal to produce annual savings to the bus company approximately equal to the anticipated annual revenue increase. Both the Commission and the Company agreed to this offer and the bus toll on the Bridge was reduced, effective March 15, 1965 from $1.00 to 13c — the same as the toll for a commute passenger vehicle.

The Bridge Board also expressed willingness to contribute to the cost of a study of a modern bus system across the Bridge from San Francisco to Marin County.

## Report of the General Manager

The theoretical capacity of the Bridge, using four lanes in the direction of peak traffic low, is 6,000 vehicles per hour. On a test day, 6,124 vehicles traveled to San Francisco during the morning peak-hour and 5,632 traveled to Marin County during the evening peak-hour. For individual toll gate lanes, the traffic volume reached more than 700 vehicles per lane, or approximately one every five seconds.

Due to the increasing volume of traffic the State of California reduced the maximum speed limit on the Waldo Approach from 65 to 55 miles per hour in the interest of reducing accidents.

Of the maintenance costs, the largest was that relating to painting the Bridge. Painting costs totaled $371,530.

Reproduction cost of the Bridge was estimated to be approximately $150,000,000 in 1965.

To supplement surveys of prior years, soundings at the north and south piers were made for the District by the U.S. Coast and Geodetic Survey. Results of the Soundings showed that there had been no significant change in the underwater topography since the prior survey and that conditions in the vicinity of the piers were stable.

To test its suitability for Bridge use, a newly developed Sodium Lucalox lamp installation was made at the north end of the Toll Plaza. Ten lamps with special fixtures were furnished by General Electric Company and installed by District personnel.

The lamp produces a high intensity, concentrated source of illumination with a yellowish color somewhat resembling the Sodium Vapor lamps that were originally installed on the Bridge and are still in use.

State Division of Highways studies were commenced for extension of Interstate Route 480 between the Broadway terminus of the Embarcadero Freeway and the Golden Gate Bridge approach in San Francisco. Prior plans of the State to widen the main San Francisco approach road through the Presidio still remain inactive due to the policy established by the Board of Supervisors of San Francisco relative to freeways in the City.

Among the other routes studied by the State was a Panhandle Freeway from the Bridge through Park-Presidio Boulevard and Golden Gate Park connecting to the Central Freeway in San Francisco. This route received endorsement from the State Highway Commission, but was rejected by the San Francisco Board of Supervisors. □

# SIXTY YEARS OF SERVICE
# Highlights from Annual Reports
## YEAR: 1965— 1966

## Report from the President

The Board of Directors of the Golden Gate Bridge and Highway District is confronted with a critical transportation problem, as the mounting traffic volume in the 1965-66 fiscal year exceeded still further the estimated practical vehicular capacity of the Golden Gate Bridge.

It is the District's public responsibility to exert every reasonable effort to force some form of effective action toward expediting the planning for a second Marin crossing, as well as investigating the most efficient and expeditious means of providing additional vehicle capacity to the existing Golden Gate Bridge.

The existing condition has been foreseen for a number of years and the Bridge District has taken all practical steps towards increasing the bridge capacity. The north end of the Bridge has been widened, the south end has also been widened recently, and a system of center-lane reversal was adopted nearly four years ago.

Further relief of substantial magnitude can only come through major alteration of the bridge or radically changing the travel habits of people. The latter holds little promise. The average occupancy of automobiles is about 1.5 persons including the driver, and group travel seems to appeal to only a very small percentage of the commuters, either by auto or by transit bus.

Getting people to give up their private cars to travel to and from work by bus, is a problem that seemingly defies solution in the immediate future.

Although mass transit is recognized as the most efficient method of travel, Marin County buses attract a relatively small number of peak hour riders. The present privately-operated bus service does not offer the degree of convenience and sufficiently low cost to induce passengers to give up the personal service features of the automobile.

The District has gone on record as favoring a comprehensive investigation of a second San Francisco-Marin crossing and was instrumental in having a new State study for this purpose initiated in the last year. Recommendations of two similar State studies in 1956-57 and 1961-62 were rejected because of public and official opposition.

It was originally estimated when the bridge was in the design stages in the late 1920's and early 1930's, that by 1970 the annual traffic would reach ten million. That figure was actually reached in 1951, now the traffic is more than 25 million a year.

Public interest demands that immediate action be taken to chart a course for meeting this crisis. A definite policy approving, in principle, a second deck on the Bridge and thus paving the way for engineering design studied was adopted by the Board of Directors in October 1965.

*[By this time it had become increasingly obvious to all Californians they lived in a state burgeoning with people, increasing wealth, and unprecedented expansion into the suburbs from central cities. Unlike many others, the District had seen the trend growing for some time as their beloved Bridge became choked with single-car commuters who crossed the span twice-daily traveling between Sonoma and Marin Counties to San Francisco.*

*This Report spares no rhetoric in stating the problem in a direct and honest manner. At a dreamy, prosperous and important time in the State's history, here was a wake-up message — a clear, fact-filled voice that was startling in its frankness, and prophetic in its call for action.]*

---

*In the rest of the World . . . .*
## HISTORICAL SNAPSHOT

**Events.** *Vietnam War escalates, includes B-52 bombing, shelling of Cambodia. Antiwar demonstrators picket White House. In Miranda vs Arizona, Supreme Court sets landmark ruling. Medicare for elderly inaugurated. Stock market dives, Dow Jones Averages hit 744.*

**Music.** *Sinatra, Strangers in the Night. Simon and Garfunkel, Sounds of Silence.*

**Books.** *Truman Capote, In Cold Blood. Yukio Mishima, The Sailor who fell from Grace with the Sea.*

**Films.** *Alfie, A Man and a Woman, A Man for All Seasons.*

**Sports.** *National and America football leagues merger into NFL. Boston Celtics win 8th consecutive championship.*

**Everyday Life.** *The word psychedelic first used to describe new street drugs. Popular new TV series included Batman, The Monkees, Star Trek.*

## 1966

A second deck involves major alterations to the bridge structure and such a project is faced with the imponderable questions of suitable approach roads that will make the probable increase in bridge capacity, and the major factor of financing.

The transportation problem on the Golden Gate Bridge is real, it is specific and it needs immediate attention. Engineering and economic feasibility studies of a second San Francisco-Marin crossing must forge ahead regardless of efforts to increase capacity of the existing Golden Gate Bridge and without delay that might conceivably result from tying this problem into the general transportation problems of the entire metropolitan area.

### Report of the General Manager

The Golden Gate Bridge once again set a new all-time annual record for traffic in its 29th year of operation, ending June 30, 1966, with 27,018,462 vehicles.

This represented an increase of 6.9% over the prior fiscal year and reflected the ever-growing population of California and the continuing upward trend in motor vehicle registration in the area served by the Bridge and the State as a whole.

The Bridge's traffic problem, just as on other major arteries of travel and in metropolitan areas, involves too many people in too many automobiles.

*[The balance of the General Manager's Report echoes the President's comments regarding the Second Deck Proposal, as well as a Second Crossing recommendation. In addition, he included comments regarding approaches, alterations to the Administration Building, the possible acquisition of some land at Ft. Baker, and installation of remote-control closed-circuit cameras to be mounted on the Bridge and connected with the Bridge Sergeant's office to aid in traffic flow and dispatch of accident vehicles and crews.]* ☐

SIXTY YEARS OF SERVICE

# Highlights from Annual Reports

YEAR: 1966— 1967

## The President's Report

The Board of Directors of the Golden Gate Bridge and Highway District has faced a crescendo of activity in the year past, marking the completion of thirty years of prudent and financially successful operation. It is well to remember that from the very day it was born, the District successfully faced conflict and turmoil, and created the *Bridge Across the Golden Gate* for the enduring use of the citizens of San Francisco and the counties to the north.

The work of these pioneers is enduring, and was conceived long before today's attitude of space flight and supersonic travel has made virtually anything seem possible, perhaps even easy. These men created, without state or federal subsidy, on the sole basis of self-reliance and bonding capacity. On the occasion of the 30th Anniversary of the opening, the Board of Directors salutes the founding principles that created the Bridge, and pledges its future stewardship to the benefit of those people who risked their properties to build it.

Now, 30 years later, the Board is confronted with a critical transportation problem that continues without letup, to exceed still further the practical capacity of the bridge. Reserve capacity of the bridge has now been exhausted, and it is imperative that some means of relief be provided to keep pace with the persistent upward trend.

Meanwhile one section of the ("*San Francisco-Marin Crossing May 1967*") report, entitled "Golden Gate Corridor," is of particular interest because it states that the transportation problems of the next 15 years can be solved most economically by a combination of a second deck, freeway system north through Marin County, freeway or bus tunnel or improved city streets in San Francisco, and a bus rapid transit system operating on exclusive bus lanes throughout. The report thus endorses the second deck project as a vital transportation link, meanwhile noting what the Board and almost everyone else knows, that is, that an entire new crossing, vehicular or otherwise, will be needed to serve the area in years to come after the second deck and related improvements are completed in four to five years.

These problems are before the Board as it begins the 31st year of operation. They are very real and without quick, pat, or flippant answers that are frequently suggested. For example, "Monorail" is not a magic answer. It is more expensive and not as fast as existing rail transit. The new generation of air cushion vehicles, as evidenced by the recently discontinued service to San Francisco International Airport, is yet too young to be practical.

Conventional rail rapid transit, which may be the Marin County solution in the 1990's is not a magic answer for today inasmuch as it requires a separate tube or bridge and needs a separate feeder system within Marin County.

---

*In the rest of the World . . . .*

### HISTORICAL SNAPSHOT

**Events.** *In what was called a long, hot summer, the U.S. was scene of the worst race riots in its history. U.S. population reaches 200-million. Three astronauts killed in launch pad fire during tests at Cape Kennedy. Six Day War between Israel and Arab States. Dr. Christiaan Barnard performs first heart transplant.*

**Music.** *Hair opens in New York. Jefferson Airplane, Somebody to Love. U.S. engineer Robert Moog creates Synthesizer. New magazine, Rolling Stone, published in San Francisco.*

**Sports.** *First Superbowl won by Green Bay, 35-10 over Kansas City.*

**Everyday Life.** *Headlines dub it "The Summer of Love." Twiggy is popular. Public Broadcasting (PBS) established.*

## 1967

On the other hand, bus rapid transit with exclusive lanes from the far reaches of Marin County to the high density of downtown San Francisco area shows great promise if it can be made attractive. This system, being studied by the Marin County Transit District, is a potential solution to the problems of the "Golden Gate Corridor."

It has long been known that for the success of any mass transit system, the California motorist must be enticed — and not forced — from his car to mass transit.

## Report of the General Manager

In its 30th year of operation, ending June 30, 1967, the bridge carried 28,350,598 vehicles. An all-time figure in annual toll revenue for the bridge was also reached, when for the first time in its 30-year history the total passed the $6,000,000 mark. A new (one-day) mark of 94,815 vehicles was established Sunday, August 7, 1966. Peak days in excess of 100,000 vehicles appear only months away.

In August 1966, the Board of Directors invited the engineering firm of Ammann & Whitney, New York, to submit a proposal for preparing preliminary plans of its concept for a second deck and total cost of the proposal. The Amman & Whitney concept was recommended by Clifford E. Paine, consultant to the District and Principal Engineer as a member of the firm of Strauss and Paine during the design and construction of the Golden Gate Bridge in the 1930's. Mr. Paine said,

"The Amman & Whitney plan with seven traffic lanes in the peak direction offers the most efficient use of the roadways without detracting from the appearance of the bridge or encroaching on its safety. This is the plan recommended for adoption."

The report covered in detail a number of solutions to the transportation problem, all of them with provision for development of mass transit facilities, either rail or bus, to serve peak commuter traffic.

Implementation of suggested alternatives in the report awaits further developments.

## Marin County Mass Transit

The Marin County Transit District received in December 1966 a comprehensive report on a feasibility study to determine economic and operational status of bus service presently performed by Western Greyhound Lines.

Recognizing the importance of Marin County mass transit to the Bridge District and the areas it serves, the Bridge Board of Directors, by formal resolution, stated the Board's intention to cooperate with the City and County of San Francisco and the Marin County Transit District to improve bus service.

Furthermore, the Board approved, in principle, the use of Bridge District funds to supplement funds raised through Marin County taxes for the improvement of inter-county bus service, it being understood that in the absence of fund commitment, City and County of San Francisco financial assistance would not be expected or needed.

*[This year's Report also included a two-page section featuring letters from the public regarding the "No charge for service" policy of the Bridge. It also reported on the success of the newly-installed closed-circuit television cameras and how they served as "a giant eye in the sky" to assist in traffic surveillance. An accompanying feature praised the courtesy of the bridge officers in the toll lanes.]* □

## SIXTY YEARS OF SERVICE
# Highlights from Annual Reports
## YEAR: 1967— 1968

### Report of the President

Each year we hear it said that the bridge has reached the limit of its capacity, yet each year sees more traffic. More than 100,000 vehicles have crossed the bridge in one day on several occasions. The record high was on Friday, June 18, 1968 when 101,335 vehicles crossed in a single day.

In May of 1967, the State of California, Division of Bay Toll Crossings published the results of a study on transportation needs between Marin and San Francisco Counties. This report stated, "The need for traffic relief in this northern portion of the Bay Area is evidenced by the present congestion, the many complaints, and the continual publicity given to the problem." The report noted the problem is primarily one of peak-period congestion. The peak is normally considered to be a 'peak hour,' but in this case it must be considered a 'peak period' since the peak now runs through two hours or more.

With respect to mass transit, the report stated, "It is obvious that an effort must be made to seek a solution of the problem of the high commuter peaks by the use of some form of mass transit."

The possibility of putting a second deck on the Golden Gate Bridge received thorough study, culminating in first a feasibility and later a financial report on how such a project could be funded. The studies indicated a second deck could be completed without marring the beauty of the bridge in any way. The world-renowned graceful exterior configuration would not change as the roadbed would be contained within the existing structure. It would be placed upon the bottom lateral bracing system installed on the bridge in 1954. With an estimated cost of approximately $50 million, and the capacity to handle vehicular traffic needs through 1985, it is easy to understand why a second deck was given much consideration.

Widespread opposition from San Francisco neighborhood groups and from elected officials of both San Francisco and Marin Counties, however, halted further second-deck action by the Board of Directors in late 1967. A resolution, adopted November 24th, stated:

> "It shall be the policy of this Board to not contract for, nor embark upon;, further studies or planning which look toward construction of a second deck until substantial improvement of mass transportation has been proved unable reasonably to relieve congestion on the Bridge during peak commute hours."

This resolution stopped further efforts of the District toward adding additional lanes of traffic across the Golden Gate. Should the public will reverse itself some time in the future, studies and plans thus far made can be removed from the shelf and dusted off for reconsideration. Because of inflation, though, the project will cost about $3 million more each year of delay.

---

*In the rest of the World . . . .*

### HISTORICAL SNAPSHOT

**Events.** *Dr. Martin Luther King and Senator Robert F. Kennedy assassinated. Richard Nixon elected President. U.S. economy threatened by inflation; prime rate rises to 6.75%.*

**Music.** *Quadraphonic sound introduced. Marvin Gaye, I Heard It Through the Grapevine.*

**Films.** *Bullitt, Butch Cassidy and the Sundance Kid, The Graduate, Planet of the Apes.*

**Sports.** *NBC cut away from a football game between NY Jets and Oakland Raiders with a minute to play. Oakland scored twice in nine seconds, won 43-32. This is the "Heidi Game."*

**Everyday Life.** *CBS anchor man Walter Cronkite denounces Vietnam War. Farm-pump maker Jacuzzi Brothers introduce Whirlpool Bath. In New York, first 9-1-1 systems begins operation.*

# 1968

### Report of the General Manager

The 1967-68 year ending June 30 was another banner year and the best on record. It is interesting to note that the 400 millionth driver crossed the bridge in November 1967. Total vehicles crossing the bridge: 30,293,793. Total operating revenue: $6,428,332.

### Toll Force Efficiency

Rush hour traffic in the toll plaza exerts great pressure on the men in the lanes, and here again, the records were broken repeatedly. One District employee has actually recorded 834 transactions per hour for his lane. This means receiving cash or commutation tickets — and occasionally making change — at the rate of once every 4 1/3 seconds. We sincerely believe the Golden Gate Bridge toll collector force to be the finest in the world.

### Maintenance

As the bridge is now 31 years old, it was deemed advisable to have a complete inspection made of the entire structure. Because of the scope of the project and the need for highly specialized technical services, the consulting firm of Ammann and Whitney was retained to make this inspection.

Inspection was underway in November, 1967, for a program that included both main tower piers, above and below the waterline, anchorages, pylons, abutments, and all concrete members. The steel inspection included both the suspended spans and approaches, anchorages, main towers, main cable wrapping, main cable wire, cable bands, suspenders, and areas of roadway slab and expansion joints.

As part of its inspection of major structural unit, Amman and Whitney will swing out and disconnect selected suspender cable pairs for testing and replacements. Fourteen suspender cables are to be disconnected to permit inspection of inaccessible parts. Three of these will be replaced with new suspender cables and the originals will be dissected and tested to confirm ultimate and fatigue strengths. ❑

**Strauss Cleaning.** *The statue of chief Engineer Joseph B. Strauss receives loving care and attention of two Bridge workers. The statue is located adjacent to the East Parking Lot. Each year thousands of visitors have their picture taken in front of Mr. Strauss, whose attention seems eternally riveted on some distant goal while his crowning achievement stands proudly behind him.*

SIXTY YEARS OF SERVICE

# Highlights from Annual Reports

YEAR: 1968— 1969

## Report of the President

Last year's annual report called attention to the never-ending increases in traffic across the Golden Gate Bridge and to predictions for heavier congestion and new one-day traffic records.

All of these have come to pass.

A total of 31,251,968 vehicles crossed the bridge during the fiscal year. A new daily high was established on Friday, July 19, 1968, with 103,647 vehicles.

As far as the commuter is concerned, the bridge is filled up.

As far as the Sunday motorist is concerned, the bridge is filled up. There is small chance of further gains in yearly traffic unless the bridge patron is willing to leave his home earlier and return later.

The outstanding action taken during the year occurred on October 19, when a one-way toll collection system was inaugurated on the Golden Gate Bridge. All vehicles passed through the lanes free northbound without stopping, and southbound tolls were doubled.

Thus, for the first time anywhere, a major toll facility achieved advantages of smooth traffic flow, safety, and reduced overhead through the collection of tolls in one direction only. They even achieved national and world-wide attention among toll bridge facilities. The 90-day test begun in October was adopted permanently on January 10. It has proven to be one of the most popular actions ever taken by the Board.

Earlier in the year, the Board commissioned Arthur Jenkins, consulting engineer and former Engineer of the District to make recommendations as to whether or not the District should implement a public transit bus system.

The "Jenkins Report" came out clearly in favor of a bus system between San Francisco and Marin-Sonoma Counties which would be designed, operated, and financed by the District from bridge revenues. Such a system, proposing 150 new buses with $9 million capital and $1.4 million yearly operating deficit, would have helped relieve bridge congestion.

Although well received, implementing action on the Jenkins Report was temporarily withheld pending completion of the report from the San Francisco-Marin Water Transportation Study Committee on the feasibility of ferries.

*[This cautionary action proved wise. Through its Committees, public hearings, personal appearances and general publicity, the Board ascertained the public will favor an improved bus transit system. As constituted at the time, the District had no authority to undertake either a rail or ferry system unless enabling legislation was passed.*

*In examining potential remedies for the growing commuter crisis, the District had asked the California Legislature the year before for such authority and it had not been granted. Now the lawmakers, for the first time, could view the "big picture" of Bay Area transportation needs before deciding on a plan of action. The Board's patience and determination proved rewarding. Perhaps much of it had to do with the next paragraph in the Annual Report.]*

---

*In the rest of the World . . . .*

## HISTORICAL SNAPSHOT

**Events.** *Richard Nixon becomes U.S. President. During final fitting, a nuclear submarine sank at dockside in San Francisco. Sen. Edward Kennedy involved in auto accident on Chappaquiddick Island, Mass. So-called People's Park established in Berkeley. Maiden flight of Boeing 747.*

**Music.** *Johnny Cash, A Boy Named Sue. The Who, Tommy. Woodstock happens.*

**Books.** *Mario Puzo, The Godfather. Kurt Vonnegut, Slaughterhouse Five.*

**Films.** *Easy Rider, Midnight Cowboy, The Wild Bunch.*

**Sports.** *Willie Mays hits 600th home run. NY Mets first expansion team to win World Series.*

**Everyday Life.** *Bank of America World Headquarters designed. Man walks on the moon, 4 of them this year.*

# 1969

## Board of Directors

There are 19 members comprising the Board of Directors. They represent the six counties that make up the District: San Francisco, Marin, Sonoma, Napa, Mendocino, and Del Norte.

Nine members represent the City and County of San Francisco. One is appointed by the mayor, four are members of the San Francisco Board of Supervisors, and four are public members appointed by the Board of Supervisors. Four represent Marin County. Two are members of the Marin County Board of Supervisors, one is an at-large member appointed by that Board, and one is nominated by the Council of Mayors and Council Members and appointed by the Board of Supervisors. Sonoma County has three representatives, one a member of the Sonoma County Board of Supervisors, one a member and nominee of the Council of Mayors and Council Members as appointed by the Board of Supervisors, and the third a public member appointed by the Board of Supervisors. Napa, Mendocino and Del Norte each have one representative selected by their respective Boards of Supervisors.

The Directors serve on standing committees which meet regularly each month. The Board meets twice a month on the second and fourth Friday at 10:00 a.m. in the Administration Building at the Toll Plaza in San Francisco. All Board and committee meetings are open to the public.

Noting the alarming increase in traffic congestion in the Golden Gate Corridor, the Board has reaffirmed its policy to assist and provide mass transit for the solution to this problem. Although the District Act now permits only bus operations, the Board supports legislation to permit the District to engage in all forms of mass transit.

While making long-range plans, the Board has meantime taken quick, effective steps to get the absolute maximum available bridge capacity working to the last percentage point.

This includes an "Automatic Vehicle Identification" system planned ultimately to serve all commute vehicles. As a prelude to "AVI," commute buses now pass through the lanes without stopping, but are identified and billed later.

The District's computer was installed in its special facility during the year. It is being programmed to yield instant traffic and revenue information, and will eventually pay for itself by reducing repairs on the bridge's original toll-lane register equipment.

During the past year, there were changes in the Board of Directors as required by new legislation increasing the membership to 18 and requiring that some of the directors also be in elective public office.

### Report of the Manager

Traffic congestion has been greatly reduced since the inception of one-way tolls. The toll-free direction now flows smoothly and even the toll-collection direction was improved as one more toll-booth lane was made available for morning commute traffic.

### Innovations

**Water Bumpers.** High energy absorbing bumpers have been installed in front of each toll booth for protection of both motorist and toll collector.

**Toll Plaza Bypass.** Improved traffic flow is expected after completion of a southbound four-lane bypass roadway off the bridge to the west of the Administration Building and onto the San Francisco approach. Preliminary design work has been started.

**Bicycle Path.** In cooperation with officials and civic groups in San Francisco, the Presidio, and Marin County, the District is providing a vital link in a bicycle path to extend from Fort Funston on the south through Marin County. This path will use the west (maintenance) bridge sidewalk on weekends and holidays. A 95-foot bicycle bridge is being built to get bicyclists from the Marin pylon to existing roads in Fort Baker enroute to Sausalito.

### Employees

Legislation enacted during the year permitted public agencies to recognize employee bargaining groups for purposes of employee salaries and working conditions. The District has recognized those unions presenting themselves as representing a majority of bridge employees in each classification. □

# SIXTY YEARS OF SERVICE
# Highlights from Annual Reports
## YEAR: 1969 — 1970

### Report of the President

This eventful year for the District saw many important changes, including the inauguration of new ferryboat service on San Francisco Bay. Concurrently, engineering and other studies of transportation in the Golden Gate Corridor were started.

These events — firsts in the District's history — occurred after the change in the District's name to GOLDEN GATE BRIDGE, HIGHWAY AND TRANSPORTATION DISTRICT, on November 10, 1969.

The legislation leading to these events became effective on that date, and had great impact on the Board's actions. The legislation requires the district to submit a comprehensive transportation plan to the California Legislature in 1971. It authorized the District to engage in all modes of transportation for the first time. Also, it restricted the District, prior to submission of the long range plan, from committing funds for a second deck on the bridge.

The legislation also requires that no more than 50% of net revenue be spent for bus transit, no more than 30% for water transit, and that the remaining 20% be accumulated or committed to long-range planning.

Studies. The major studies undertaken as a result of the legislation were:

A long-range transportation facilities plan to serve the Golden Gate Corridor until the year 2015.

Studies to focus on community values, socio-economic impact, and environmental consequences of the plan being considered.

Design of the new Marin County-San Francisco ferryboat system.

Supporting studies of financing methods, visual identity symbol, and logotype for all transportation activities.

Transit Operations. November 10, 1969, also saw the inauguration of ferryboat service by the Bridge District with commute runs between the Ferry Building, San Francisco, and Tiburon. The Board authorized a 90-day trial period using the Motor Vessel "Harbor Emperor" at a time when normal service was halted due to a strike. Action by the District was possible through the cooperation of the vessel's owner, the labor unions involved, and officials and citizens of Tiburon and Belvedere.

---

*In the rest of the World . . . .*
## HISTORICAL SNAPSHOT

*The Bridge was conceived and built during an important, expanding time in history. It's birth was difficult; it exists in a real world. These are a few of the events, near and far, that happened concurrently with its development and growth. As you'll see, from time to time events occurring far from San Francisco Bay held influence over the business of the Golden Gate Bridge.*

**Events.** *Environmental Protection Agency established. The Apollo 13 lunar mission was near-disaster. Bill banning cigarette advertising on radio/tv signed, effective 1/1/71. Amtrak takes over running of passenger trains in U.S. Intel introduces the microprocessor. The New York Times publishes The Pentagon Papers. Four students killed when National Guard troops fire on war protestors.*

**Music.** *Shostakovich, Symphony No. 15. Godspell and Jesus Christ Superstar open. Marvin Gaye, What's Going On? Janis Joplin, Pearl. The Rolling Stones, Brown Sugar.*

**Deaths.** *Coco Chanel, Igor Stravinsky, Louis Armstrong, Nikita Khrushchev.*

**Films.** *The Conformist, dir Bertolucci; Five Easy Pieces, Jack Nicholson; Love Story, starring Ali MacGraw, Ryan O'Neil; M*A*S*H, dir Robert Altman; Performance, dir Nicholas Roeg.*

**Sports.** *Curt Flood files antitrust suit against organized baseball. Wimbledon women's doubles won by Billie Jean King and Rosemary Casals for third time in four years. Top golfing moneywiners, Lee Trevino, $157,037 and Kathy Whitworth, $30,235. The America's Cup was defended by U.S. yacht, Intrepid, defeating Britain's Gretel II, four races to one.*

**Everyday Life.** *Inflation and 5.8% unemployment brought faltering economy in U.S. Designers pushed the Midi, but youth fashions inspired tie-dyed garments, granny dresses and secondhand military attire. The Golden Gate Bridge District changed its name and expanded into areas the founders never imagined.*

## 1970

Service to Tiburon ended April 5, 1970, when the strike was over and the vessel returned to the owner. During the 147-day operating period, the District carried over 150,000 passengers.

The District committed itself to ferryboat operation on a permanent basis in June with the delivery of a 113-foot twin diesel 596-capacity vessel for commuter service between San Francisco and Sausalito. Renamed the "Golden Gate," the vessel is the first of an expected fleet.

Emphasis was given to water transportation on San Francisco Bay in 1969 by the report prepared by Arthur D. Little, Inc. The report concluded that ferryboats should be returned to the Bay and suggested such unconventional craft as hydro-foils and air cushion vehicles for extended northern Marin County runs and conventional hull ferryboats for shorter runs. The report recommended that the Bridge District be the single agency in charge of the integrated bridge-bus-ferry system.

Legislation was introduced during the year (that) would create a Metropolitan Transportation Commission. The bill would require the Bridge District to make its transit plans in the Golden Gate Corridor conform to Commission's regional plan.

The District's Board of Directors has recognized the legislative mandate to enter all forms of transportation within the Corridor, and has noted the heavy capital outlays involved. The Board took action by raising the price of a 25-ride commute book from $6.50 to $10.00 beginning with the January 1970 book. This action is expected to add $1 million annually to toll revenues received.

## Report of the General Manager

Total vehicles crossing the bridge total 32,725,026 for the fiscal year. Total operating revenue exceeded $7 million for the first time, $7,409,360.

The present daily crush of 29,400 commuters is estimated to reach 47,000 by 1980. The problem continues to be the one-man one-car custom of the commuter. A check of 17,000 vehicles during the morning peak on March 10, revealed 71% contained the driver only.

**Maintenance.** A major engineering inspection reported the Golden Gate Bridge to be in "excellent condition." Although some repairs and replacements should be instituted as soon as possible, the report continued, there are no defects so critical that immediate emergency steps must be taken.

The estimated cost of repairs totaled $10½ million over the next decade, including $7 million to replace suspender ropes. Other items noted were repairing roadway, tightening bolts in cable bands, repair of floor beams and repairs of bottoms of towers and bolts which anchor the main cables.

**Water Transit.** The five month ferryboat operation was of great value in gaining operating experience (and) helped to alleviate peak period bridge congestion and clearly demonstrated that service of this kind was desired by the public.

**Land Transit.** In September, the General Manager recommended to the Board a $7.7 million program to initiate a commuter bus service. A ferryboat system in northern Sausalito by 1972 was also urged as a part of a balanced transportation system.

**Personnel.** The total number of employees within the District has increased only 100% since the bridge was opened to traffic in 1937, while traffic is up 1000%. In January of 1970 there were 216 District employees. Statistics do not include ferryboat personnel. □

## SIXTY YEARS OF SERVICE

# Highlights from Annual Reports

## YEAR:1970 — 1971

[This Report featured a full-color rendition of the new logo, a three-element design. The cover also included three photographs: A Golden Gate Transit bus; a north tower view of traffic on the Bridge, and a side view of the Sausalito ferry.

Since the first year of the Golden Gate Bridge, Annual Reports of the District have dealt with the problems, opportunities, and financial results of Bridge operation. This year's Report emphasizes the expanding responsibilities of the Golden Gate Bridge, Highway and Transportation District. As portrayed by the design and photographs on the cover, the District is now concerned with all modes of transportation in the Golden Gate Corridor. The green in the logotype design suggests surface transportation over the green hills of the northern counties; the blue signifies the District's commitment to increasing ferry travel on the waters of San Francisco Bay; the bright orange represents the famous span of the Golden Gate Bridge. This unique design heralds a new era of responsibility for the Golden Gate Bridge, Highway and Transportation District.]

### Report of the President

The year ending June 30, 1971 was one of the most challenging faced by the Board of Directors since the early days.

(The Board) moved forward on plans to establish a Golden Gate Bus Transit System to provide comprehensive service into San Francisco from Santa Rosa and Marin County.

(It) began operation of the M.V. Golden Gate, bringing back ferry service between San Francisco and Sausalito, and took steps to create a ferry system serving other Marin points.

(We) completed and submitted to the State Legislature a long range plan to provide guidelines for the eventual creation of a rapid transit system in the Golden Gate Corridor.

(We) paid off the last of the bonds issued to finance the construction of the Golden Gate Bridge.

In 1930, during the depths of the depression, the people of the District voted more than three to one to issue $35 million in bonds to build the Golden Gate Bridge. This was accomplished in spite of strident opposition by those who said the Bridge could not be built, and if built, would not stand.

Now that the debt has been retired there have been questions concerning the District's toll policy. **Contrary to popular belief, the original legislation creating the District did not call for abolition of tolls when the structure was paid for.**

Instead, it was noted that the management should re-evaluate the toll structure when that day arrived, recognizing that funds would continue to be needed for operations, maintenance and repair. What could not be visualized was the massive population growth that would one day choke traffic on the Bridge during peak periods, requiring new solutions to solve the transportation problems of those living in the Golden Gate Bridge Corridor.

Philip Spaulding, a prominent naval architect who supervised modification of the M.V. Golden Gate, was retained to develop plans for a more extensive ferry system. He recommended service be established between San Francisco and the Larkspur-Corte Madera area, and eventually to a northern Marin point above San Rafael.

The Board of Directors agreed in principal to the purchase of new 165-foot, 750 passenger vessels to be designed by Spaulding.

Plans have also moved ahead rapidly for the beginning of a modern bus system to serve commuters and off-peak riders between Santa Rosa and San Francisco.

The Marin Transit District had developed what was called the "Optimum Bus Plan." It was designed to carry Marin residents from the near vicinity of their homes to downtown San Francisco, unlike the present Greyhound service which essentially operates along Highway 101. The Marin Transit District asked the Golden Gate Bridge, Highway and Transportation District to implement the plan and the Directors agreed.

---

*In the rest of the World . . . .*

## HISTORICAL SNAPSHOT

**Events.** *In the first foreign trip ever undertaken by a Japanese Emperor, Hirohito met with President Nixon at Anchorage, Alaska. Swiss scientist develops computer language, Pascal.*

**Music.** *Follies, Godspell, Jesus Christ Superstar open.*

**Films.** *A Clockwork Orange, Dirty Harry, The French Connection, The Godfather, The Last Picture Show.*

**Sports**. *U.S. table tennis team visits China. Washington Senators (AL) become the Texas Rangers. Bobby Fisher becomes first American to qualify for World Chess Championship.*

**Everyday Life.** *Disney World opens in Orlando, Florida. The dismantled London Bridge is reassembled at Lake Havasu City, Arizona. Big fashion statement: Hot Pants.*

## 1971

Further studies were completed leading to the award to General Motors, after competitive bid, of a contract for 132 air conditioned buses with sound repression and pollution control packages.

While bus and ferry plans were being pursued, the District was developing a long-range transportation facilities plan. (Consultants) reviewed all past proposals for rapid transit service between San Francisco and Marin. With the aid of a Citizens Advisory Panel, these were narrowed to the five most promising possibilities.

The five alternate plans were taken to the public for their review in a series of 21 public hearings throughout the District. The consensus was that a flexible system should be created There was concern that a fixed guideway system might hasten the urbanization of areas served by the system which are now largely suburban.

The District therefore, recommended to the Legislature the phased development of a rapid transit system. The Golden Gate Bus System, now being implemented, would grow in stages into a rapid transit system operating on its own rights-of-way. Route alignments and way structures would be designed so that a fixed guideway system could evolve when and if such a transition is proved desirable.

Buses, ferries, important preliminary plans for mass rapid transit — all are moving toward the ultimate goal: reduction of the average 92,000 vehicles, most carrying a lone passenger, that daily cross the Golden Gate Bridge.

The Bridge suffers not from too little success, but rather too much. It is the Board of Directors' continuing challenge to move people quickly through the Golden Gate Corridor, not merely by the Bridge alone, but by all other means to meet the transportation needs of the present and the future.

## General Manager's Message

A total of 33,497,080 vehicles crossed the Bridge during the period, a record. Peak traffic day was Sunday, August 2, 1970, when 109,152 vehicles crossed the Bridge. Toll revenues were $7,978,284.

Projections indicated that the present 30,000 commuters daily will reach 47,000 by 1980. Car pools are represented by only about one car in 50. The existing 60 feet of Bridge roadway (six 10-foot lanes) cannot be widened, and therefore other solutions must be found. We believe they are provided in the Golden Gate Bus Transit System and Golden Gate Ferry System.

**Maintenance.** Replacement of worn rivets and high-strength tension bolts, underway since July 1969, is now half completed. Eight men work full time on this project.

Sandblasting and repainting the entire bridge with an inorganic zinc silicate paint, begun in 1968, will continue for another five years. When completed, this new primer coat is expected to last for 20 to 25 years, with repainting in the interim only for the sake of appearance, rather than as a necessity to protect the span. This is a slow and difficult job. For example, sandblasting and repainting the arch-span over Fort Point alone took two years to complete.

Before the end of 1971, work should begin on replacement of the vertical suspender cables that support the bridge deck. The first phase of the project, to replace about half the cables, will take 18 months to complete.

Preliminary testing has been done this year to improve lighting on the Bridge. Further tests will be undertaken in the next fiscal year.

The automatic vehicle identification research program was completed and the system is now operational. Transponders have been placed in District vehicles to record passage through the toll gates. The equipment will be installed in District-operated buses, and eventually may be available to record passage of commuter automobiles so that driver could be billed monthly.

**Golden Gate Bus Transit.** The service is designed to bring commuters from Santa Rosa and Marin County to within a short walk of their San Francisco destinations. Presently some 3,500 people commute to San Francisco by Greyhound bus. With the advent of the Golden Gate Transit System, the number of bus commuters is expected to increase to at least 6,500 within a relatively short time. This will remove approximately 2,200 private vehicles from the Bridge during the commute periods.

**Golden Gate Ferry Transit.** The District's single ferry, M.V. Golden Gate, operating between San Francisco and Sausalito, carried 716,000 passengers during its ten-and-one-half months of operation in the past fiscal year. Patronage was steadily mounting and it was anticipated that approximately 900,000 people will have been carried by the vessel at the end of its first year of operation, August 14, 1971.

As noted in the President's report, additional vessels have been designed for the District's ferry service. It is hoped that the first of the several new vessels can be built and placed in service early in 1973.

In the meantime, the District is seeking federal approval to undertake a test of hovercraft vessels to determine if they are suitable for passenger use on San Francisco Bay. Manufacturers of this type of vessel have offered to provide them on a loan basis at nominal cost to the District so the tests can be made under actual operating conditions. □

# ATTENTION READERS — Oops!
### Publisher's Mistake Is Your Gain

Sharp-eyed readers may have discovered by now that in the vicinity of page 180, something is missing. Maybe you didn't catch it. We didn't — until it was too late in our production schedule.

OK, we'll confess — what's missing is the Annual Report for the Fiscal Year, 1947-1948. Yes, the whole darn thing. This omission is being corrected and subsequent editions of *THE BRIDGE — A Celebration* will include it. In case you don't remember, this was the year 'Scrabble' came on the market, *South Pacific* won a Pulitzer Prize, Citation won the Triple Crown, and the term "Cold War" was coined. It was a meaningful year for the Golden Gate Bridge, too.

We don't know how this happened (yes, we do), but it has been suggested that as a result of our blunder, we may have created a genuine collector's item. We believed our First Edition would be purchased and saved by both book and Bridge aficionados — people who collect First Editions because their value will increase over time, and people who love the Bridge and want every scrap of information they can gather about this wonderful structure.

If this is true, our bungle is your gain because we're going to correct our error and print a new edition of *THE BRIDGE*, and replace all copies at retail outlets. However, interested collectors and Bridge buffs may order an *original edition* (with the mistake) directly from the publisher at the address below.

**NOTE:** Obviously, if you're reading this, you already own a Collector's Copy. Congratulations. For you lucky readers, this is the only opportunity available to those of you who may want to share your good fortune with family or friends. If you order, we'll include a separately printed "addendum" for the missing year to make your copy complete.

This offer is for individuals only. Absolutely no dealers will be honored. As an accommodation, we'll pay the five-dollar (each copy) shipping and handling cost. Copies are $19.95. All funds in US $. In order that as many collectors and Bridge lovers as possible can share in this offer, please limit your order to no more than three copies. (We have no way to check on this, so it's the honor system.)

Send your order to: **OOPS BOOK OFFER. 72 Locust Avenue, Mill Valley, CA 94941-2131.** Personal Check, Money Order, VISA or MasterCard accepted. Orders with credit card number, expiration date and signature may be faxed to (415) 383-5685. California residents, please add $1.45 sales tax for each copy. All copies will be in perfect condition.

Orders shipped within 24 working hours of receipt. Please allow about three weeks for domestic delivery. Orders filled on a first come-first served and this offer is limited to copies on hand.

THE PUBLISHER

SIXTY YEARS OF SERVICE

# Highlights from Annual Reports

YEAR: 1971 — 1972

## Report of the President

On July 1, 1971, the first day of the new (fiscal) year, the District celebrated retirement of the last of the bonds to finance construction of the Golden Gate Bridge.

*[The bonds were non-callable. The District used its accumulated reserves to apply to renewal and construction projects for the Bridge, as well as for matching funds to apply against federal grants to establish a bus system and to expand ferry operations.]*

During the many public hearings we held on our consultants' studies and reports, we found that there was great reluctance by the public to endorse a rapid transit system in the Golden Gate Corridor. While population projections of the various planning agencies indicated that future growth would require such a system, many doubted the projections. They expressed the belief that such a rapid transit system, which could cost as much as $987 million to build, would in itself attract the increased population.

Accordingly, the Board of Directors of the District adopted as policy that there should be a phased development of transit. First, a modern bus system would be created which would operate on existing streets and highways. The existing single ferry boat would be expanded by the addition of three new high-speed, 750-passenger ferries. Then, studies would be undertaken to define private rights-of-way for the buses, first in San Francisco and then in Marin and Sonoma Counties.

The private rights-of-way would be planned so that they could accommodate fixed guideway rapid transit, if and when this became necessary to satisfy transportation needs.

The (state) Legislature accepted this policy and requested the District to further refine its long-range plan and return in 1973 with more details.

The bus system went on stream in December of 1971, with the service first being provided within the County of Marin. On January 3, 1972, trans-bay service was implemented.

Plans for the expanded ferry system also moved forward. While (the Board) hoped that the vessels could be built in the San Francisco Bay Area, the low bidder was Campbell Industries, Inc., of San Diego. Cost of the three new boats and the terminal facilities in Larkspur and Sausalito in Marin County, and in San Francisco, is projected at $24,717,405.

The first of the new boats is scheduled for delivery in June of 1974, and the two remaining vessels at four-month intervals thereafter.

With delivery of the last ferry boat, the Golden Gate Bridge, Highway and Transportation District will have a fully-paid-for integrated bus and ferry transportation system valued at some $36 million. It should be noted, too, that the Golden Gate Bridge, now valued at some $200 million, is also owned free and clear by the people of the six counties of (the) District.

---

*In the rest of the World . . . .*

## HISTORICAL SNAPSHOT

**Events.** *Watergate affair begins with burglary arrest June 17. Supreme Court rules death penalty illegal. Dow Jones hits 1000 in November. Screening of luggage on U.S. airlines made mandatory. CAT scans introduced.*

**Music.** *Grease, on Broadway. Roberta Flack, The First Time Ever I Saw Your Face. Helen Reddy, I Am Woman.*

**Radio.** *In Britain, final broadcast of The Goon Show*

**Films.** *Cabaret, dir Bob Fosse; The Concert for Bangladesh; The Discreet Charm of the Bourgeoisie, dir Luis Bunuel; Last Tango in Paris, starring Brando; Play It again, Sam, dir and starring Woody Allen.*

**Sports.** *Summer Olympics in Munich marred by Arab terrorist attack on Israeli athletes; Mark Spitz wins 7 gold medals, all world records. Baseball ruled exempt from antitrust law, ending Curt Flood case.*

**Everyday Life.** *Dallas Cowboys become first football team to introduce professional cheerleaders. First video game, "Pong" introduced.*

# 1972

## General Manager's Message

A record-breaking 34,435,542 vehicles crossed the Golden Gate Bridge during the 1971-72 fiscal year. The peak traffic day was Sunday, August 8, 1971, when 113,756 vehicles crossed the span. Total revenues: $8,382,867.

**Golden Gate Bus Transit:** Within three months the trans-bay buses were carrying up to 6,000 people each way during commute hours. This was an increase of 1,500 passengers over the number of commuters carried by Greyhound Lines in the months immediately previous to the start of Golden Gate Transit service.

By August 1972, Golden Gate Transit was operating 175 buses. The rapid acceptance of the Bridge District's bus system by the public leads us to project capacity operations during commute hours by early 1973. This means that we hope to be transporting more then 7,000 people daily in and out of San Francisco during commute hours.

Also, Golden Gate Transit provides service within Marin County on weekdays. Cost of this local service is subsidized by the Marin County Transit District and patronage is growing monthly.

**Golden Gate Ferry System.** In two years of operation, the M.V. Golden Gate carried nearly two million people between Sausalito and San Francisco, and patronage continues high. With the arrival of new vessels (in 1974) service will be extended to include Larkspur in central Marin County, where a 25-acre ferry terminal site has been purchased.

Currently, the Bridge is carrying almost three million vehicles monthly. The M.V. Golden Gate is carrying an average of 91,000 passengers a month. Some 345,000 people are riding Golden Gate Transit buses on trans-bay routes monthly, and another 122,000 are riding between Marin County points.

**Bridge Operations.** In a joint effort with the California Department of Highways, an exclusive lane for buses traveling northbound on Highway 101 over Waldo Grade in Marin County during the evening commute hours has been established.

This exclusive lane, which became operational in September, is provided by taking two lanes from the southbound traffic. These are separated by plastic tubes inserted in the pavement, following the same system that has been used for a number of years to provide reverse lanes on the Golden Gate Bridge. One of the two lanes is used by the buses, and the other acts as a buffer to separate the bus lane from oncoming traffic.

If the experiment proves successful, it is expected that an exclusive bus lane can be established during the morning commute hours as well.

Work to replace the lighting fixtures on the Bridge is expected to be completed in December of 1972. The type of bulb originally used is no longer being produced, and this necessitated installation of the new fixture. The new lights will produce the familiar golden glow which has become a Golden Gate Bridge trademark, but will provide illumination that is several time the intensity of the original system.

The project to repaint the entire Bridge with an inorganic zinc silicate paint has continued on schedule. Begun in 1968, this project will continue for another four years. The new paint is expected to last for 20 years with minor maintenance.

**Personnel.** The District now has 567 people in its employ. The major increase in employees has been in the Bus Division, which now has total employment of 304.

The Ferry Division now has 29 people in its organization. No major increase in staffing is contemplated until early in 1974.

*[The expanding District activities brought about a reorganization of management during the year. Three separate Divisions were created: The Bridge Division, The Bus Division, and the Ferry Division, each with its own manager and staff. The General Manager remains the chief operating officer with general supervision of each of the newly-created Divisions.]* □

SIXTY YEARS OF SERVICE

# Highlights from Annual Reports

YEAR: 1972 — 1973

## President's Message

The development of an integrated bus and ferry transportation system to alleviate traffic congestion in the Golden Gate Corridor, on the Golden Gate Bridge, and on the streets of San Francisco, moved forward as planned during the fiscal year ended June 30, 1973.

On August 21, 1972, the U.S. Department of Transportation, Urban Mass Transportation Administration (UMTA)[Now the Federal Transportation Administration, FTA], approved a grant of $16,478,100 to pay two-thirds of the capital cost of the District's expanded ferry system.

The District's present ferry, the M.V. Golden Gate will continue to operate between San Francisco and Sausalito as it has since 1970. The (vessel) has been carrying an average of one million passengers a year.

Expansion of the ferry service is expected to attract millions of passengers annually, since the new Larkspur terminal will be located in an area readily accessible to more than half of Marin County's commuter population.

**Bus System's Success.** The Golden Gate Transit Bus System, which began service at the start of 1972, is now operating at near-capacity during commute hours. Almost 8,000 people are riding Golden Gate Transit buses to and from work every day. Previously, some 4,300 people had commuted on the Greyhound system which the District replaced.

No small factors in this success are the experience, competence and courtesy of our bus drivers, as the numerous letters of commendation received by the District testify.

**Benefits to Public.** To give a total picture of the benefits of the integrated bus and ferry transportation, it should be noted that in 1971, before the start of the bus system, Greyhound was carrying some four million passengers a year in the Golden Gate

Corridor. In this past fiscal year, Golden Gate Transit Bus and Ferry Systems carried about seven million passengers, and are expected to carry eight million next year.

This moving of people by mass transit, rather than by automobile, has resulted in a marked slowing in growth of auto traffic on the Golden Gate Bridge, with resulting conservation of energy, reduction in air pollution and traffic congestion on the streets of San Francisco, and thus the enhancement of the Bay Area environment.

On a cost-benefit basis, mass transit is far cheaper for the community at large than the massive capital and operation costs that support the use of the private automobile.

**Long-Range Corridor Planning.** While the development of our integrated mass transit systems has occupied much of the Board of Directors' attention during the past fiscal year, the Golden Gate Bridge has not been neglected.

---

*In the rest of the World . . . .*

## HISTORICAL SNAPSHOT

**Events.** *Wounded Knee, S.D. occupied by American Indian Movement (AIM). Watergate scandal saturates U.S., indictments begin. Agnew resigns, replaced by Gerald Ford. Inflation hits 8.5%. OPEC embargo on oil, long gas lines, factories close. Vietnam peace agreement signed in Paris.*

**Deaths.** *Lyndon Johnson, Noel Coward, Picasso, John Ford, J RR Tolkien, W H Auden, Pablo Casals, David Ben-Gurion.*

**Music.** *Sondheim, A Little Night Music; Roberta Flack, Killing Me Softly With His Song: Elton John, Goodbye Yellow Brick Road; Stevie Wonder, I Believe.*

**Everyday Life.** *Barcodes first used in supermarkets. Commuters flock to public transit. Endangered Species Act passed. Kojak a television hit. With development of urethane wheel, Skateboarding becomes popular in U.S.*

## 1973

The program to replace about half of the suspender cables on the Bridge is moving on schedule. New roadway lights have been installed which provide several times the intensity of the original lights installed 36 years ago. Priorities have been established for other work which, along with the regularly scheduled maintenance programs, will continue to keep the span in essentially new condition.

With encouragement from the public, and with their commitment to use transit to help support its costs, the Golden Gate Bridge, Highway and Transportation District will continue to do its part to provide reliable, effective alternatives to the private automobile.

### General Manager's Message

A total of 34,620,920 vehicles crossed the span in the fiscal year ending June 30, 1973. This represented an increase of 185,378 vehicles (over) the previous fiscal year. For the moment, the continuing trend in traffic increases has been slowed. Average daily traffic was 94,852. The annual toll revenue was $8,668,900.

**Golden Gate Bus Transit.** In its first year and a half of operation, the Golden Gate Transit Bus System has been operated from temporary headquarters. However, construction of the first phase of the main terminal in San Rafael has been completed, and the full complex will be finished in the current fiscal year. Satellite facilities in Santa Rosa have also been completed, and planning is underway for the Novato terminal.

The exclusive permit bus lane over Waldo Grade, operated effectively during the year, and its benefits in speeding bus commuters home in the evening prompted the California State Division of Highways to extend the lane for nearly a mile. The exclusive lane, which operates on the "wrong side" of the highway median over Waldo Grade and crosses to the correct side just beyond the grade, now is approximately 5.5 miles long and may be extended further as improvements to U.S. 101 are made in Marin County.

**Golden Gate Ferry Transit.** Construction was started on the first vessel on December 12, 1972. By the close of the fiscal year, it was about 25 per cent complete, with construction on a second vessel well along. When all three vessels are in service, they are expected to carry some 3,200 commuters between Marin County and San Francisco during the peak morning hours every weekday.

Planning for the main terminal and base of operations for the Ferry Transit Division in Larkspur has been mainly completed, and a contract has been awarded for a comprehensive environmental impact report. New terminals are also planned in Sausalito and in San Francisco. These are necessary because passengers will enter and leave the new vessels by the upper deck, rather than the lower deck used by the District's single ferry, the M.V. Golden Gate.

**Golden Gate Bridge Division.** The rivet replacement project — and the program to completely repaint the Bridge with long- lasting, inorganic zinc silicate paint — continued on schedule during the year. Both of these programs, designed to reduce maintenance costs over the long term, will take several more years to complete.

Also moving ahead on schedule is the replacement of suspender ropes which support the roadway beneath the main cables. The first contract was for replacement of 254 of the 500 ropes. Replacement of the remaining ropes will be determined upon completion of current studies.

The Bridge District's tow service, formerly limited to the Bridge and its immediate approaches, has been extended to cover Waldo Grade for the length of the permit bus lane and to serve Doyle Drive.

The Division put grooves in the roadway surface (of Doyle Drive) to provide improved traction in wet weather. Another innovation was the installation of signs asking motorists to turn on headlights while driving on the approaches during daylight hours. □

SIXTY YEARS OF SERVICE

# Highlights from Annual Reports

## YEAR: 1973 — 1974

### General Manager's Message

Traffic across the Golden Gate Bridge showed the first annual decrease since the gas rationing days of World War II. A total of 32,824,780 vehicles crossed the span, a 5.19 percent decrease from the 34,620,920 vehicles recorded the previous year, and approximately the same number of vehicles carried in 1969-70.

While the impact of the fuel shortage was undoubtedly the major cause of the traffic decline, the Golden Gate Transit operations can be credited as well. The District's buses and ferry transported approximately 6,838,000 trans-bay passengers, and an additional 2,838,000 bus patrons traveled between points in Marin and Sonoma Counties.

Bridge operating revenues totaled $9,491,837, an increase of 9.49% over the prior year, attributable principally to the 25-cent increase in the roundtrip auto toll which became effective March 1.

**Golden Gate Bus Transit.** The annual toll was affected significantly by the fuel shortage, which peaked in March of 1974. While commute-hour auto traffic on the Golden Gate Bridge declined to 1969 levels, for the first time the District's bus fleet experienced a large number of standees. With their automobiles, and morning commute patronage dropped from more than 9,600 back to the 8,500 range, which was still considerably above what had been carried during the morning commute hours the year before.

The District was operating during the period with a fleet of 188 coaches. An additional 32 buses are slated for delivery before the end of the current calendar year, and another 30 have been ordered for delivery in the spring of 1975.

**Golden Gate Ferry Transit.** The District's single ferry, the M.V. Golden Gate, continued to ply the Bay on a regular schedule between San Francisco and Sausalito, and again registered a million-passenger year.

Both the bus and ferry systems were severely affected, from a cost standpoint, by the escalating prices of diesel fuel and other supplies and equipment. Yet, both the bus and ferry system completed the year under budget.

**Golden Gate Bridge Operations.** On the Golden Gate Bridge, the rivet replacement program and the repainting of the Bridge with long-lasting, inorganic, zinc silicate paint, continued as scheduled.

In response to the fuel crisis, it was decided to use only half the roadway lights on the Bridge during clear weather. With the new lighting system installed during the previous year, it was found that using only half the fixtures provided almost as much light as the old sodium-vapor lamps that were replaced. The District embarked upon, and is continuing, other energy-saving measures.

A new program to create a buffer lane between north and southbound traffic during mid-day and night-time hours, in order to reduce the potential for head-on collisions, was inaugurated and proven successful.

*[Once again events on the opposite side of the world had direct, measurable effects on the Golden Gate Bridge. This time it was the Persian Gulf and the Arab Nations' oil boycott which created shortages and sent gasoline prices soaring throughout the U.S. It also sent millions of Bay Area commuters scurrying to mass transit, as lines at the gas pumps grew longer, shortages more acute. During the fiscal year, almost two million fewer vehicles crossed the Bridge. Where did those commuters go? This year's Report tells the story. Because the General Manager's Message is richer in detail, we have reversed the order of presentation this time.]*

---

## In the rest of the World . . . .
### HISTORICAL SNAPSHOT

**Events.** *President Nixon resigns, result of Watergate incidents. Gerald Ford assumes presidency, pardons Nixon of any crimes he may have committed. Freedom of Information Act passed. Economic crisis. Auto sales down 20%, housing starts down 40%. Patricia Hearst kidnapped.*

**Films.** *Chinatown, Godfather II, Papillon. 700-thousand videos rented in U.S.*

**Sports.** *Henry Aaron breaks Babe Ruth home run record. Muhammad Ali defeats George Foreman in Zaire title bout. Little League announces its teams open to girls. Research poll shows 33,900,00 Americans playing tennis.*

**Books.** *Erica Jong, Fear of Flying; John LeCarre, Tinker, Tailor, Soldier, Spy; Alexander Solzhenitsyn expelled from USSR following publication of The Gulag Archipelago.*

**Everyday Life.** *"Streaking" has brief burst of popularity.*

## 1974

The exclusive bus lane over Waldo Grade in Marin County, which operates on the "reverse" side of the median strip during evening commute hours, proved very successful in the past year. This program has enabled Golden Gate Transit to expedite schedules with considerable cost savings.

## President's Message

Expansion of the Golden Gate Bridge, Highway and Transportation District's mass transportation services continued during the past fiscal year. Through grants by the Urban Mass Transportation Administration (UMTA) which, as of the date of this report, total $37.5 million, the District has been able to plan, build, and purchase transportation facilities and equipment having an original cost in excess of $53 million.

**Integrated Bus-Ferry System.** Upon completion of our transit program as it is presently planned, the District will have in service 250 buses, three new high-speed ferry boats, and the ferry M.V. Golden Gate, which the District has been operating between Marin and San Francisco Counties since August of 1970.

**Bus Service Patronage.** We have been extremely pleased over the acceptance of the Golden Gate Transit bus service. Load factors during the commute period regularly exceed 90 percent.

**High-Speed Ferry Program.** Presently, the Larkspur Transit Terminal is planned for completion early in 1976. Feeder-bus routes will provide service to the new terminal, time-coordinated to meet the 30-minute departure schedules of each high-speed ferry. The first of the three new ferries was launched on October 5, 1974. Delivery is scheduled to the District on San Francisco Bay early in 1975. The first vessel, named the G.T. SAN FRANCISCO ("G. T." stands for gas turbine), will go into service soon after its arrival.

**Long-Range Transit Plans.** If these transit programs had not been implemented, traffic congestion would by now be causing significant back-ups of automobiles on the Waldo Grade in Marin each morning, and in the evening on Doyle Drive and Lombard Street in San Francisco's Marina District.

Population projections have indicated that without mass transportation service in the Golden Gate Corridor, there would be an eventual need for 18 lanes of freeway through Marin, another bridge at the Golden Gate, and a second deck on the Golden Gate Bridge, with more freeways, auto congestion and air pollution within San Francisco. The effects of an auto-only transportation solution for the Golden Gate Corridor are clearly unacceptable.

**Transit System Investment.** The capital cost of creating the Golden Gate Bridge District's transit system will, when completed, be in excess of $53 million, with a cost to the District of about $16 million.

The capital improvement and replacement program on the Golden Gate Bridge have continued at a cost in the last fiscal year alone of about $2 million. When improvement programs on the Bridge have been completed, the Bridge will be in essentially new condition, having a replacement value of approximately $300 million.

**Transit Costs, Fares and Subsidies.** The problem faced by the District, as well as by every other public transportation agency in the nation, is meeting the on-going costs of the transit operations. If fares are charged at a rate that equals operating costs, travel by transit may become too expensive. Fares, therefore, must be adjusted to levels which are attractive to riders.

This cost consideration requires support from other sources. The "other source" most often relied upon by public transit operations is the property tax. The Golden Gate Bridge, Highway and Transportation District, however, has no taxing powers and must, therefore, operate basically within its own income.

The District has been, and is, dedicated to providing tangible, meaningful public services. It is meeting the complex challenges created by the problems of traffic congestion in the Golden Gate Corridor with a balanced, integrated transportation system of buses and ferries.

We intend to continue our commitment to provide efficient and effective transportation services for the benefit of the people of the Golden Gate Corridor as well as our entire Bay Region. □

SIXTY YEARS OF SERVICE

# Highlights from Annual Reports

YEAR: 1974 — 1975

## President's Message

On September 1, 1975, the Golden Gate Bridge, Highway and Transportation District submitted to the State Legislature, as requested, a report on its long-range transportation planning. Titled, "Golden Gate Corridor Transportation Facilities Plan, Phase II," the report detailed the District's activities in all phases of its operations from the time it made its last report to the Legislature on April 1, 1971, through the fiscal year ended June 30, 1975.

For this reason, the District's Annual Report for the year just past is taking this reduced form.

The total vehicles crossing the Golden Gate Bridge, 33,238,810. For Golden Gate Transit, the number of bus passengers for the year was 7,885,470; the number of ferry passengers, 1.087,804. ☐

*[The report indicated in the brief notes on this page required the time and attention of the Staff and Board of Directors. As they state, the comprehensive plan supplied most of the information usually related in The President's Message and the Report of the General Manager in the annual report.*

*The details of this report and actions taken are revealed in subsequent Annual Reports, highlights of which are included hereafter.]*

---

*In the rest of the World . . . .*

## HISTORICAL SNAPSHOT

**Events.** *Two assassination attempts on President Ford, one in Sacramento by Squeaky Fromme, another in San Francisco by Sara Jane Moore. After 19 months, Patricia Hearst captured in San Francisco apartment. Former Teamster president James R. Hoffa reported missing. Microsoft founded by Bill Gates, 19, and a friend.*

**Films.** *Jaws, Nashville, One Flew Over the Cuckoo's Nest.*

**Television.** *New shows: Saturday Night Live, Starsky and Hutch, Wheel of Fortune.*

**Sports.** *World Football League fails. Junko Tabei of Japan first woman to climb Mt. Everest.*

**Everyday Life.** *First "drive-thru" McDonalds opens. Over one-million "Pet Rocks" sold in U.S.*

## 1975

[Once again the Report carried only the financials and a brief letter from the President. Despite the brevity of the President's "To the People of the District" one-page report, it was a year that had more than its share of problems, operationally and fiscally.]

## SIXTY YEARS OF SERVICE

# Highlights from Annual Reports

## YEAR: 1975 — 1976

### President's Letter

On September 1, 1975, we submitted to the State Legislature a report on the second phase of the long-range planning efforts to provide transportation improvement in the Golden Gate Corridor. The studies were primarily related to the movement of people from the south end of the Bridge into downtown San Francisco.

The studies indicated that through 1985 improvements made to accommodate Golden Gate Transit buses should be on surface streets with preferential treatment using contra-flow and reverse lanes. After 1985, it was indicated some exclusive rights-of-way may be necessary. The report noted that additional commuter capacity might be obtained after that date through expansion of ferry service.

Start-up of the District's expanded ferry transit service from Larkspur was again delayed. The new terminal was ready for service at the end of the fiscal year but additional work was required on the three new vessels to correct deficiencies in their propulsion systems. The District will not accept the vessels until they meet all specifications.

Golden Gate Transit bus service was interrupted for 64 days in a labor dispute with drivers beginning on April 12. Expected massive traffic jams did not occur, however, since commuters moved into car pools and adjusted their travel times.

During this period, the District eliminated bridge tolls for car pools of three or more as an emergency measure, and this policy has been continued. Car pools now average more than 1600 daily, about twice the pre-strike number.

The four-year job of replacing all the suspender ropes on the Golden Gate Bridge has been completed at a cost of more than $8 million. Eventually the concrete roadway on the Bridge will require replacement, but the present surface will remain serviceable for another decade, consultants report.

As you will note, (in the financial report, not included here), the District ended the fiscal year with total revenues exceeding expenses. However, as costs continue to increase, it becomes obvious that an adjustment in transit fares and the bridge toll will soon be necessary. The Board of Directors is continuing to monitor the situation closely.

In closing, I would like to express appreciation for the public's support of our efforts to reduce auto congestion by providing effective transportation systems at reasonable costs to the rider. With the public's help we have been very successful to date, and we hope to continue this record.

Total vehicles across the Golden Gate Bridge in fiscal year, 1975-76, 34,871,856. There were 7,046,797 bus passengers; 1,103,810 ferry passengers. □

---

*In the rest of the World . . . .*

## HISTORICAL SNAPSHOT

**Events.** *Production of convertibles by U.S. auto makers ends. Concorde supersonic jet service begins. Ford and Carter debate on TV. Genentech, world's first genetic engineering firm founded. Apple Computers founded.*

**Music.** *Musical group, ABBA Sweden's biggest export earner after Volvo.*

**Books.** *Alex Haley, Roots.*

**Sports.** *Dorothy Hamill emerges as star of Winter Olympics, Innsbruck. Reserve clause in baseball allows players to become free agents after 5 years. Richest baseball contract signed, Reggie Jackson, NY, $2.9 million, 5 years.*

**Films.** *Rocky, Network, Marathon Man.*

**Everyday Life.** *Legionnaire's disease discovered in Philadelphia. Barbara Walters leaves NBC for ABC. The Muppet Show begins.*

# 1976

SIXTY YEARS OF SERVICE

# Highlights from Annual Reports

YEAR: 1976— 1977

## President's Letter

To the People of the District:

This is a brief report to you on the activities of the District during the fiscal year which ended June 30, 1977.

The major event of the year was undoubtedly the dedication of the Larkspur Ferry Terminal and the entry into service of the first of our long-awaited new high-speed commuter ferry boats, the G.T. Marin. This occurred on December 11, 1976. The second new vessel, the G.T. Sonoma, was put into service on March 7, 1977.

The two new ferries, and the G.T. San Francisco, which will soon be added to the operating fleet, will enable us to continue to hold the commute-period traffic counts on the Golden Gate Bridge at the 1969 level.

The added transit capacity is solely needed, as our buses continue to run full during commute hours, and the number of commuters in the Golden Gate Corridor are increasing at an accelerated rate. Where in recent years we would expect to add 1,000 commuters annually, recent county indicated that last year's increase exceeded 2,000. Without the transit option, these new commuters would be added to the auto traffic on U.S. 101, the Bridge and the streets of San Francisco.

The greatest challenge facing the District is to obtain the funds necessary to continue its transit service. Unlike other transit organizations in this area, as well as most others elsewhere, Golden Gate Bus and Ferry Transit receives no local tax support. We must meet essentially all of our operating costs from current income, and this is exceedingly difficult with continuing inflation.

Therefore, tolls on the Golden Gate Bridge, which not only cover repair and maintenance of the bridge but the necessary subsidies for transit operations, must be increased. Transit fares also need to be adjusted upward, and it is expected that this will be accomplished in the new fiscal year. Without these necessary increases, it will be necessary to cut back on transit services, since without recourse to the tax rolls, the District simply will not be able to operate transit at necessary levels.

It is the Board of Director's goal to meet at least half of the transit costs from the fare box. The remaining monies come from bridge tolls not needed for maintenance and remedial work on the bridge and from state and federal operating assistance funds.

Last year, as you will note, the District had a deficit of almost $1 million. This was covered by reserves. However, without toll and fare adjustments our modest necessary reserve funds will vanish. We know that this is not in the public's interest, and we trust that the residents of the District, and those who use its services, will support our efforts to operate in a fiscally responsible manner.

In 1976-77, 35,378,544 vehicles crossed the Golden Gate Bridge. Bus passengers totaled 8,374,579 and ferry patronage rose to 1,708,321. □

---

*In the rest of the World . . . .*

## HISTORICAL SNAPSHOT

**Events.** *President Jimmy Carter pardons Vietnam-era draft resisters. Huge power failure strikes New York, 3776 looters arrested. First National Women's Conference held, Houston. Ban on saccharin proposed by FDA. Trans-Alaska pipeline opens. Minimum wage raised to $3.35/hr. In New York, two men diagnosed with AIDS.*

**Music.** *Elvis dies at home from drug overdose. Bing Crosby dies.*

**Films.** *Annie Hall, Close Encounters of the Third Kind, Star Wars, Saturday Night Fever.*

**Television.** *Broadcast of mini-series, Roots, captures ratings.*

**Sports.** *Oakland defeats Minnesota in Superbowl XI, 32-14. Pele begins play with U.S. soccer team.*

**Everyday Life.** *First U.S. made diesel automobiles are sold. Larry Flynt convicted of promoting obscenity.*

**1977**

SIXTY YEARS OF SERVICE

# Highlights from Annual Reports

YEAR: 1977— 1978

### President's Letter

At the end of the fiscal year, the last major element of our water transit program was completed with the opening of the San Francisco Ferry Terminal. This new facility is now accommodating up to 9,000 passengers daily on the Larkspur and Sausalito ferry runs. We expect, with our facilities now complete, to see continuing increases in ferry patronage.

In the first week of this new fiscal year, on July 5, 1978, the Golden Gage Bridge experienced its highest traffic day in history, when 118,796 vehicles crossed the span. This exceeds the previous record of August 8, 1971, by more than 5,000 vehicles, and dramatizes the continuing growth of auto traffic in the Golden Gate Corridor.

Despite the District's developments of improved transit, and the numbers of people riding in carpools and vanpools, every month traffic on the bridge increases up to three percent above the same month the year before. An average of more than 100,000 vehicles now cross the bridge daily, while a few years ago the daily average was about 90,000.

Right now, the 20,000 autos that cross the bridge during morning commuter hours carry an average of 1.5 persons per car. If that average were increased to two persons, it would greatly help in accommodating present and future growth in the Corridor. To this end we are working with other public agencies and local governments to encourage carpooling, and we solicit the interest and support of all the people of the District.

Again, we want to acknowledge the efforts of all the District's personnel in the Bridge, Bus and Ferry Divisions in providing courteous and conscientious service to the public. We appreciate their interest and enthusiasm.

In the 1977-78 fiscal year, 36,031,236 vehicles crossed the Bridge. Bus passengers totaled 8,647,031, while 2,142,448 rode the ferries. □

---

*In the rest of the World . . . .*

## HISTORICAL SNAPSHOT

**Events.** *Availability of energy major problem for U.S. government and industry. The Love Canal area of Niagara Falls, NY declared a disaster area. Mass suicide by Rev. Jim Jones, followers, in Guyana. Mayor George Moscone, Supervisor Harvey Milk, shot to death in San Francisco City Hall. First legal casino outside Nevada opened in Atlantic City.*

**Books.** *John Irving, The World According to Garp. Armistead Maupin, Tales of the City.*

**Sports.** *WBA Boxing: Ali loses, then regains title from Leon Spinks.*

**Everyday Life.** *Jogging craze heats up in U.S. Sneakers account for 50% of shoe sales in U.S. Disco culture reaches peak.*

## 1978

SIXTY YEARS OF SERVICE

# Highlights from Annual Reports

YEAR: 1978— 1979

## President's Letter

### To The People of the District

During the year, a number of important events took place which had significant effects on the District's financial position.

The fuel shortage, plus escalating fuel costs, had a dramatic effect on District operations. In July of 1978, the beginning of our fiscal year, diesel fuel was selling at an average of 35 cent per gallon. By the end of June 1979, the price had risen to an average of 61 cents per gallon, and was headed even higher.

The excess of revenue achieved in 1978-79 will be used to help meet the increased operating costs. These higher fuel costs, together with high rates of inflation, are depleting the District's unrestricted reserves. In the future, it will be necessary to explore every available source of revenue, if we are to be able to continue transit operations at present levels.

The fuel situation had a direct impact on bus and ferry patronage, as people gave up their cars in favor of public transit. Bus patronage set a record in May, when 9,802 persons were carried on transbay buses during the morning commute period. The previous record occurred during the 1974 fuel shortage, when 9,668 persons were carried.

Ferry patronage had been increasing gradually during the year. When the fuel crunch came, a record was set in May, when 53,271 passengers were carried during a one-week period.

Bridge traffic showed a drop during the fuel crisis. Prior to the fuel crunch, daily traffic averaged 96,538. During the height of the fuel shortage, traffic was down to 93,054. This severe drop in Bridge traffic resulted in lower Bridge revenues.

One costly financial question was resolved during the year. This involved a 40-million dollar price tag for replacement of the Bridge roadway. The U.S.

Government has agreed to fund 80 percent of the replacement cost.

Total vehicles crossing the Bridge for the fiscal year 1978-79, 36,723,760. The number of bus passengers grew to 8,745,609. Passengers using the ferries totaled 2,059,908. ☐

*In the rest of the World . . . .*

## HISTORICAL SNAPSHOT

**Events.** *Ayatollah Khomeini returns to Iran from Paris exile. Consumer Price Index up 13.3%. Use of petroleum products declines. Nuclear near-disaster occurs at Three Mile Island, Pa. Department of Education created as 13th Cabinet-level agency. Hostages taken at U.S. embassy in Teheran.*

**Music.** *Biggest pop music groups: Abba, Pink Floyd, The Police, Village People.*

**Deaths.** *Nelson Rockefeller, Jean Renoir, Mary Pickford, John Wayne, Richard Rodgers.*

**Sports.** *Muhammad Ali retires. U.S. Open Tennis singles won by John McEnroe, Tracy Austin, at 16 youngest player to win women's single.*

**Everyday Life.** *"Palimony" case between actor Lee Marvin, former companion, Michelle Triola. First spreadsheet program for personal computers. Laser video disks go on sale.*

## 1979

# THESE ARE THE DATES

The Golden Gate Bridge, as you've seen and will continue to see, has a rich history. Real history isn't merely dates, but the time when things happened are important. These are Highlights.

**1872** Earliest discussion of a bridge across the Gate.

**1916** Newspaper editor revives notion of a Gate bridge.

**1918** SF Board of Supervisors seeks feasibility study.

**Jun 28, 1921** Strauss submits sketches, cost estimate of $27 million to SF City Engineer Michael O'Shaughnessy. Other estimates range $60 to $77 million.

**Jan 13, 1923** SF and North Bay counties meet, form Association of Bridging the Gate; draft legislation to form Golden Gate Bridge and Highway District.

**May 25, 1923** The Association influences the State Legislature to legally form the District.

**Dec 20, 1924** U.S. Secretary of War issues provisional construction permit to build the Bridge.

**Dec 4, 1928** The District is incorporated to design, construct and finance the Golden Gate Bridge.

**Jan 23, 1929** District holds first meeting in SF.

**Aug 15, 1929** District Board appoints Strauss Chief Engineer.

**Aug 11, 1930** U.S. War Department issues final permit for construction of the Bridge.

**Aug 27, 1930** Strauss submits his final plans to District.

**Nov 4, 1930** Voters of six counties approve $35 million bond issue to finance the Bridge.

**Jan 5, 1933** Construction begins.

**May 27, 1937** Bridge opens to pedestrians.

**May 28, 1937** FDR presses telegraph key to announce the opening of the Golden Gate Bridge to the world - ahead of schedule, under budget.

**Dec 1, 1951** Violent winds close Bridge for nearly three hours. Subsequently District installs bottom lateral bracing that increases roadway torsional rigidity 35 times.

**Jun 21, 1956** 150 millionth vehicle crosses the Bridge.

**1957** Series of earthquakes strike SF Bay Area, registering as high as 5.5 on the Richter scale. No damage to the Bridge, but a few widows break at the Toll Plaza.

**May 27, 1962** The Bridge celebrates its 25th Anniversary with a civic luncheon and a parade of antique cars.

**Oct 29, 1963** Reversible lanes inaugurated on the Bridge, greatly aiding peak traffic flow.

**Oct 19, 1968** World's first one-way toll systems begins on the Golden Gate Bridge.

**Nov 10, 1969** In response to mounting traffic congestion on the Bridge, California Legislature directs the District to develop a mass transportation system for the Golden Gate Corridor. The word "Transportation" added to District's name.

**Jun 1970** District takes delivery of the M.V. (Motor Vessel) Point Loma after complete overhaul to ready her for ferry service between Sausalito and San Francisco.

**Aug 15, 1970** Sausalito Ferry Terminal dedicated; newly-christened M.V. Golden Gate begins service to SF.

**Dec 15, 1970** "Ferry Feeder Bus Service" to Sausalito Ferry begins with leased Greyhound buses.

**Jul 1, 1971** Remaining original bonds for construct of the Bridge are retired. $35 million in principal and nearly $39 million in interest have been financed entirely from Bridge tolls.

**Sep 1971** General Motors delivers the first 20 of a total of 132 new coaches.

**Dec 15, 1971** Local intracountry bus service begins serving Marin County.

**Jan 3, 1972** Golden Gate Bus Transit transbay commuter service begins with 132 buses.

**Jan 31, 1975** Construction of Golden Gate Bus Transit's state-of-the-art bus maintenance facility completed, in San Rafael.

**Apr 1976** District initiates toll-free Bridge passage for vehicles with 3 or more occupants during peak commute.

**Dec 11, 1976** Larkspur Ferry Terminal dedicated; first new ferry, G.T. (Gas Turbine) Marin, begins service to San Francisco.

**Mar 7, 1977** Second vessel, the G.T. Sonoma, is added to Lakspur run.

**Sep 1977** Third vessel, G.T. San Francisco, delivered. Two ferries provide daily service, third serves as alternate.

**Jun 17, 1978** San Francisco Golden Gate Ferry Terminal dedicated.

**Mar 1982** Sixteen new GM coaches replace old buses. New buses able to "kneel" and provide built-in wheelchair lifts.

**Dec 1983** General Motors delivers 51 additional Advanced Design coaches.

**Dec 1983** First G.T. ferry travels to San Diego for conversion from gas turbine water jet propulsion to diesel engines with twin propellers; reduces fuel costs by 60% annually.

**Feb 22, 1985** One billionth vehicle crosses the Bridge.

**Nov 17, 1985** G.T. Ferries returned from conversions are rechristened M.S. (Motor Ship). First time all three vessels operate schedule between SF and Larkspur.

**Aug 15, 1985** Construction completed to replace original Bridge roadway with modern orthotropic steel plate deck.

**May 24, 1987** The Golden Gate Bridge celebrates its 50th Anniversary. Gift Center opens in historic "Roundhouse."

**Jul, 1987** Bus Transit accepts delivery of 21 new MCI (Motor Coach Industries) buses with innovative wheelchair lifts and higher passenger capacity.

**Jan, 1990** Bus Transit receives 80 new TMC (Transportation Manufacturing Corp.) Coaches, each with passenger lift and room for two wheelchairs.

**Nov 9, 1990** Ground breaking ceremony for the C. Paul Bettini San Rafael Transit Center. Bettini former San Rafael mayor, GGBHTD Board member.

**Mar 3, 1991** Commute service between Marin and Sonoma Counties begins to relieve inter-county congestion on Hwy 101. First such service in Northern Corridor.

**Sep 1991** Bus Transit accepts delivery of 63 new TMC coaches, each with passenger life and wheelchair accommodation.

**Jan 1, 1992** Bus Transit celebrates 20th Anniversary of transbay service, began Jan 3, 1972.

**Jan 12, 1992** C. Paul Bettini San Rafael Transit Center opens serving Golden Gate Bus Transit, Greyhound, Marin Airporter, Santa Rosa Airporter, WhistleStop Wheels and area taxi services.

**Aug 1992** Santa Rosa Bus Facility dedicated as the Helen Putnam Transit Center, honoring Putnam for her dedication to serving North Bay communities, and District.

**Aug 28, 1993** Contracts awarded for Golden Gate Bridge seismic retrofit final design work.

**Mar 7, 1993** Three-year demonstration bus service begins between Marin County, East Bay via San Rafael-Richmond Bridge, funded by Metropolitan Transportation Commission.

**Nov 1, 1993** Golden Gate Transit intercounty paratransit service begins.

**Jan 28, 1994** Contractors complete pedestrian railing replacement on the Bridge.

**Sep 1994** Bus Transit receives 45 new flexible coaches featuring innovative front-door wheelchair lifts and 45 passenger seats.

**Aug 15, 1994** Ferry Transit completes 25 years of service.

**Jun 7, 1995** Northwestern Pacific Railroad Authority (NWPRA) formed to hold title and preserve portion of the Northwester Pacific Railroad between Novato in Marin County and Healdsburg in Sonoma County; and Novato to Lombard in Napa County.

**Apr 29, 1996** Purchase of 139 miles of Northwestern Pacific Railroad right-of-way completed.

**Sep 13, 1996** Golden Gate Bridge designated as a double-fine zone to aid enforcement of 45 MPH speed limit.

**Jan 1, 1997** Bus Transit celebrates 25th Anniversary.

**May 27, 1997** 60th Anniversary of The Golden Gate Bridge.

SIXTY YEARS OF SERVICE

# Highlights from Annual Reports

YEAR: 1979— 1980

## President's Message

Fiscal Year 1979-80 was a busy, challenging, and exciting year. Work continued toward keeping the Golden Gate Bridge in the best possible condition, while at the same time keeping Bridge auto traffic at manageable levels.

Despite significant increases in the number of commuters entering San Francisco each morning from the Golden Gate Corridor (42,000 at last count), Bridge commute auto traffic has been held to about the 1970 level. We are encouraged by the fact that total vehicle traffic cross the Bridge was down by over 3/4 million vehicles from the previous year; that ridership on Golden Gate buses increased by more that 700,000; and that carpools, vanpools and club bus ridership all showed corresponding increases.

On the 43rd anniversary of the building of the Golden Gate Bridge, planning efforts were continued for the redecking of the Golden Gate Bridge and replacing the sidewalks. Engineering and design work will take place during 1981, using funds allocated from the Federal Highway Trust Fund. In addition, seismic upgrading plans for Bridge approaches were completed which will give additional protection against damage from earthquakes. Also, new toll booths are being designed with a prototype booth to be tested at an early date.

We were pleased with the results of a research study sponsored by the Regional Transit Association which disclosed that Golden Gate Transit Services received the highest rating by both rider and non-riders of any transit operator in the Bay Area.

Among the District's efforts to gain new revenue for transit this fiscal year, were increased transit fares (bus and ferry) and reductions in free carpool hours. However, we will be facing deficits in the next fiscal year. Since the 1/2 cent sales tax for transit failed on the Marin County ballot in April, the District will be looking at the possibility of higher Bridge tolls and transit fares.

The Board remains dedicated to maintaining the Golden Gate Bridge, furnishing transit services, and offering incentives to alternative methods of Ridesharing, in the most efficient and cost effective manner.

The Golden Gate Bridge carried 35,531,296 vehicles during the year. Golden Gate Transit had 9,472,688 passengers; while there were 1,117,508 ferry passengers, 361,257 Club Bus riders and 192,500 Vanpool riders. □

---

*In the rest of the World . . . .*

## HISTORICAL SNAPSHOT

**Events.** *Hostage rescue mission in Iran fails. Mt. St. Helens erupts in southwest Washington state. Summer-long heat-wave in U.S. hits hard in 20 states. Sales of U.S. autos at 19-year low. Inflation 12.4%. Ford Motor Co. loses $595- million in third-quarter. Sale of personal computers climbs. Pres. Carter orders boycott of Summer Olympics in Moscow.*

**Music.** *Barnum, Les Miserables on Broadway. David Bowie, Ashes to Ashes. Sony introduces portable tape player, the Walkman. John Lennon shot outside his apartment building in New York by former mental patient.*

**Deaths.** *Cecil Beaton, Jesse Owens, Jean-Paul Satre, Alfred Hitchcock, Peter Sellers, Colonel Sanders.*

**Films.** *Elephant Man, The Empire Strikes Back, Kagemusha, Ordinary People, Raging Bull, The Shining.*

**Television.** *Newest critical success, Hill Street Blues.*

**Sports.** *At Winter Olympics, Lake Placid, NY, U.S. hockey team scored major upset over favored USSR team. NCAA basketball championship, Louisville defeats UCLA, 59-54. In the World Series, Philadelphia defeats Kansas City, 4 games to 2.*

**Everyday Life.** *Ronald W. Reagan elected president. Television episode of evening soap-opera, Dallas, Who Shot J.R.?, achieves highest rating in television history.*

**The Golden Gate Bridge** *found inflation and rising energy costs affecting Bridge and Transit services.*

## 1980

SIXTY YEARS OF SERVICE

# Highlights from Annual Reports

YEAR: 1980— 1981

## The President's Message
## To The Constituents of the District:

This past year has been an interesting, challenging, and busy one. We have been most pleased with the progress made on the essential Bridge re-decking project, and expect to solicit bids during the coming year on this crucial project.

Transit and Ridesharing services have continued to hold Bridge auto commuter traffic down.

During the year, it became quite clear that with the impacts of inflation and higher energy costs, the District could not continue to provide transit and Ridesharing services unless Bridge tolls and transit fares were raised.

Studies prepared by the staff covering the five year period from FY 1980-81 through FY 1984-85 indicated substantial and increasing annual deficits with the then-existing tolls and fares. After a series of hearings on a number of alternative toll and fare proposals, the Board adopted a Bridge auto toll of

$1.25 effective March 1, 1981, and raised inter-county bus fares by 16%. Bridge tolls for the handicapped remained at 50 cents, and for the first time, motorcycles were allowed to cross the Bridge free of tolls during the 6-9 AM and 4-6 PM weekday commute periods. In addition, some local bus fares and the Larkspur Ferry fares were raised.

Following implementation of the $1.25 toll on March 1, delays were experienced at the toll plaza and the Board authorized the sale of discounted commute book tickets. With growing concern over traffic delays, the Board, in June, ordered a suspension of the $1.25 toll and directed a test of $1.00 tolls to measure the time difference required to collect the extra 25 cents. This test was to take effect in July.

Even though gasoline prices continued to climb, this fact did not seem to discourage automobile drivers. A total of 36,393,422 vehicles crossed the Bridge, compared to 35,531,296 during FY 1979-80.

In response to the public request for bus service out the 19th Avenue corridor in San Francisco, Route 66 service was started in June. Plans are also being pursued to establish a Geary Boulevard Golden Gate Bridge Service.

In order to effect economies, the Board approved a diesel conversion program for the Larkspur ferries, and in response to the public's concerns, initiated plans to reduce the Larkspur Ferry service until such time as the new diesel engines are installed.

Effective June 15, 1981, an austere ferry schedule was implemented which reduced midday runs and weekend runs along with some commute schedules. These cutbacks will conserve fuel and maintenance costs.

The 10 millionth ferry passenger was appropriately honored and designated on October 30, 1980. ☐

---

*In the rest of the World . . . .*

## HISTORICAL SNAPSHOT

**Events.** *Iranian hostages released moments after Ronald Reagan sworn in as president. Economy falters, inflation at 14%. Sandra Day O'Connor becomes first woman on Supreme Court. Risk of coronary death linked to cholesterol.*

**Books.** *Martin Cruz Smith, Gorky Park. Minoru Oda, Hiroshima. Paul Theroux, The Mosquito Coast.*

**Science.** *World's longest suspension bridge over of River Humber in Britain. IBM launches personal computer, using MS-DOS. French railways introduce high-speed Train a Grande Vitesse.*

**Everyday Life.** *U.S. Center for Disease Control connects AIDS with HIV virus. Cost of medical care rises 12.5%.*

## 1981

# SIXTY YEARS OF SERVICE

# Highlights from Annual Reports

## YEAR: 1981— 1982

*[Since it's formation, the District had met a long series of challenges that changed with the times. They ranged from the social and political to the environmental and financial. In each instance, foresight and planning had always been hallmarks of the District's management style. But how do you plan for disaster? Directors, Managers and Employees were subjected to a startling new test — this one from Nature itself.*

*In this Report, the President understates the District's heroic actions during an emergency situation that is remembered to this day as "The Storm of '82."]*

## Report of the President

Fiscal Year 1982 presented many challenges to the District, some of them man-made and some caused by nature, such as the disastrous storm of January 1982.

At the start of the Fiscal Year we experimented with a variety of tolls in an effort to find a toll scheme which would bring in required revenues while keeping toll plaza traffic congestion to a minimum.

A number of public hearings were held to obtain viewpoints of Bridge patrons to the various toll schemes. Public response indicated a split toll, $1.00 Sunday through Thursday and $2.00 Friday and Saturday was the most acceptable.

The great storm of January 1982 had a major impact on Bridge, Bus and Ferry operation. As a result of slides and flooding in Marin and Sonoma Counties, our bus transit operation was brought to a standstill and the Bridge was closed to traffic at various periods due to mud slides on the Waldo Grade to the north of the Bridge.

The Golden Gate Larkspur ferries and Sausalito ferry became literally the only means of transportation to and from Marin and Sonoma Counties. All three Larkspur ferries were pressed into service and several private ferries were chartered. One day alone, a record 12,275 passengers were carried on the Larkspur service.

Golden Gate buses were used to rescue many persons who were stranded in their homes during the storm. The District's Ridesharing Division Personnel worked long hours in setting up vanpooling and carpooling.

As a result of efforts made to provide transportation during this difficult period, the District was the recipient of a 1982 Transportation Award from the Metropolitan Transportation Commission.

The first step in a $56.3 million Bridge deck replacement project was initiated on the Golden Gate Bridge that replaced three 15' x 50' prototype concrete deck section with prefabricated orthotropic steel deck sections. The first section was installed at night in a 9-hour experiment. The section "fit like a glove," confirming that the project was feasible, and traffic could be accommodated with very little inconvenience to the public.

Steps were taken to increase seat availability on Golden Gate Transit buses. A retrofit program was introduced which increased the number of seats from 45 to 49. This was equivalent to adding 18 additional buses without the associated cost. The Bus Division also took delivery of 16 new General Motors advanced design buses equipped with wheelchair lifts.

Golden Gate Bus Transit celebrated its 10th Anniversary of service on January 8, 1982. Since 1972 District buses have carried more than 95 million passengers.

---

*In the rest of the World . . . .*

# HISTORICAL SNAPSHOT

**Events.** *Unemployment hits 10%. Largest Gross National Product decrease since 1946, but recession appears over at year-end. AT&T agrees to break up Bell System.*

**Music.** *CD players go on sale. Michael Jackson, Thriller. U2, New Year's Day.*

**Films.** *Bladerunner, Diner, ET, Gandhi, Tootsie.*

**Sports.** *At World Cup in Spain, Italy defeats W.Germany, 3-1 in finals. Super Bowl XVI, San Francisco beats Cincinnati 26-21.*

**Everyday Life.** *Kodak introduces the disc camera. IBM launches 3084 computer. Computer games smash hit; TIME Magazine features "Pac-Man" as Man-of-the-Year.*

# 1982

The District received Federal Grant approval in the amount of $2.1 million towards the diesel conversion of two of the Larkspur Ferry vessels. It is anticipated that when this conversion is completed, there will be a 60% savings in Ferry fuel costs.

The Ridesharing Division continues its efforts to encourage carpoolintg and vanpooling. Twenty-one new vans were purchased to be used in the District's Vanpool demonstration program. A commuter (lease) carpool program was authorized as well as a flex-pool demonstration service. This backup van service provides flexibility to ridesharers who occasionally miss their regular rides.

In summary, this has been a busy and challenging year. All of the efforts and projects directed for maintaining the Bridge, keeping the flow of traffic at a manageable level, and furnishing a cost-effective transit system, could not have been accomplished without the dedicated efforts of the Board, employees, and the cooperation of the public. We thank you all for your efforts.

Total vehicles on the Golden Gate Bridge: 35,752,524. Golden Gate Transit carried 10,688,401 bus passengers and 1,274,671 ferry passengers. There were 395,630 Club Bus riders and 91,646 Vanpool riders. □

**Follow the Line.** *This photo was taken from the north end of the Bridge during the height of The Great Storm of '82. Notice the absence of traffic, save for a District vehicle on an inspection tour. However, look very carefully at the white divider line down the center of the roadway. As it approaches the top of the naturally-arched deck, it veers strongly left (east) as the roadway is blown in that direction by high winds. The storm caused mud slides on Waldo Grade which also closed traffic enroute to the Bridge.*

# SIXTY YEARS OF SERVICE

# Highlights from Annual Reports

## YEAR: 1982— 1983

### President's Message

Fiscal year 1982/83 was a remarkable and eventful year for the Golden Gate Bridge, Highway and Transportation District a milestone year during which the District:

- Awarded a construction contract to replace the original roadway and sidewalks on the Golden Gate Bridge.
- Acquired one mile of abandoned Northwestern Pacific Railroad right-of-way from Corte Madera Creek to southern San Rafael.
- Purchased a 1.5 acre parcel in Larkspur Landing, adjacent to the railroad right-of-way.
- Reached agreement with the City of San Rafael Redevelopment Agency on selection of a site for the proposed San Rafael Transportation Center.
- Implemented important bus service and route changes in Marin and Sonoma Counties.
- Award a contract for the conversion of the District's three Larkspur ferries from gas turbine waterjet to diesel twin-screw propulsion.

**Traffic and Safety Improvement Task Force.** The safety of motorists cross the Golden Gate Bridge is of paramount concern to the District. A Task Force, comprised of District Directors and staff, representatives from Caltrans, and the California Highway Patrol was established to:

- Review speed limit regulations, the adequacy of enforcement, and recommend applicable regulatory changes.
- Review and evaluate the feasibility of a movable barrier.
- Examine the feasibility and structural capacity of the Bridge to accommodate a second deck including cost estimates and consideration of legislative needs.

**Northwestern Pacific Railroad Right-of-Way.** District policy calls for key sections of the Northwestern Pacific Railroad Right-of-Way parallel to U.S. 101 to be acquired for public transit purposes, as it becomes available, from Corte Madera to Highway 37 in Novato.

In the summer of 1982, Southern Pacific announced its intention to abandon 2.5 miles of the right-of-way in central Marin. As lead agency for the Northwestern Pacific Railroad Interagency Task Force, the District moved ahead, and in February 1983 acquired one mile of the right-of-way between southern San Rafael and Corte Madera Creek, with an option to purchase an additional 1.5 miles south of Corte Madera Creek.

**San Rafael Transportation Center Site.** The year 1983 saw the conclusion of an eight year effort to select a site for the proposed San Rafael Transportation Center that is mutually acceptable to the City of San Rafael, the County of Marin, and the District. The site will offer greater passenger convenience, lower transit operating costs, and adjacency to the Northwestern Pacific Railroad right-of-way.

---

*In the rest of the World . . . .*

## HISTORICAL SNAPSHOT

**Events.** *Census Bureau announces poverty rate highest in 18 years. Auto production increased 10.2%; new home construction 60%. Hospital costs rose 10.7%. Sally Ride first female astronaut in space. GM/Toyota auto venture approved, plant in Fremont, CA.*

**Sports.** *For first time since organized in 1870, U.S. loses America's Cup yacht race to Australia. Baseball top spectator sport in U.S., 78,051,343 fans.*

**Science.** *IBM introduces PC with built-in hard drive. Apple introduces pull-down menu, mouse controller. Researchers isolate AIDS virus.*

**Everyday Life.** *Auto accident deaths lowest in 20 years; attributed to use of safety belts. Leg warmers top fashion item. Consumer craze creates shortage of Cabbage Patch dolls.*

## 1983

## Marin and Sonoma County Bus Service Changes.

Faced with large deficits, the Marin County Transit District asked the Bridge District to assume operation of local bus service in Marin County for a one-year trial period commencing July 1, 1983. In exchange, the MCTD will transfer to the Bridge District its revenues for that period. Concurrently, Marin County basic transbay and local bus service was restructured to reduce service costs to match the revenue from the County.

The Sonoma County Board of Supervisors requested the District to withdraw local bus service in Sonoma, so that the County itself could provide local service.

Public hearings were held in Marin and Sonoma Counties by all involved agencies before implementing service and route changes.

Commemoration of the Start of Bridge Construction. The Golden Gate Bridge celebrated the 50th anniversary of the start of its construction on January 5, 1933. Commemoration of this date attracted international media attention. Several thousand people viewed an exhibit of historical photos, memorabilia, and contemporary children's art relation to Bridge construction.

## General Manager's Report

**Bridge Traffic.** It is evident that the saturation point for morning and evening commute traffic is close at hand.

Traffic on the Bridge increased 3.46% during fiscal year 1982-83 to a record total of 36,990,346 vehicles. On June 13, 1983, 117,560 cars crossed the span, the second highest day since the Bridge opened. On many days, over 7,200 vehicles passed through the toll plaza during the peak 7-8 am period.

The single passenger vehicle count increased steadily, averaging 16,500 per day. Correspondingly, commute bus patronage decreased by approximately 1,000 passengers a day a situation not unique to Golden Gate Transit. The weak 1982 economy and declining fuel prices contributed to a nationwide drop in the use of public transit.

**Bridge Closed Due to High Winds.** On December 22, 1982, for the second time in its history, high winds were responsible for the closure of the Golden Gate Bridge to all traffic.

The hazard of light trucks and vans being blown over by the 68 mph winds caused the Bridge to be closed down from 4:21 pm to 6:00 pm. The Bridge functioned as anticipated by the engineers and suffered no damage from the winds.

With the Bridge closed, the Ferry Transit Division assumed the responsibility of transporting northbound commuters. The Division ferried 3,549 passengers, triple the average evening patronage.

**Bridge Maintenance and Roadway Replacement.** On November 12, 1982, a construction contract for the Golden Gate Bridge Deck and Sidewalk Replacement Project was awarded to Dillingham-Tokola, of Pleasanton, California, for $52,495,000. The Federal Highway Administration is funding approximately 80% of the total project costs of $61,095,000.

Seismic upgrade of both the San Francisco and Marin approaches to the bridge was completed in September of 1982. Underwater inspection of the Bridge piers and fenders was also completed. Consultants reported that the piers are sound and their structural integrity is not in question.

**Bus Transit Fleet Expansion.** Contracts were awarded to General Motors Corporation to purchase 51 suburban type Advanced Design coaches for $10.1 million. All new buses will be equipped to accommodate wheelchairs. At the San Rafael Bus Facility, design is underway to expand the administration building and the maintenance shop.

**Diesel Conversion of Larkspur Ferries.** The long awaited contract to convert the District's three 725-passenger aluminum vessels from gas turbine-waterjet to diesel twin-screw propulsion was awarded during 1983. The total contract bid by Southwest Marine of San Diego, California was $4,990,310.

**Ridesharing Programs.** The year 1983 marked the beginning of new promotion strategies for Ridesharing. Employer outreach programs encouraged North Bay firms to form company vanpools, and quarterly toll booth handouts promoted Ridesharing to single occupant vehicle commuters. A Home-End marketing campaign to North Bay households generated 1,565 Ridesharing applicants.

As a result, fifty new Golden Gate Vanpool groups were formed. In the North Bay, 186 Golden Gate Ridesharing owner-operated and private company vanpools carried over 2,000 commuters a day. □

SIXTY YEARS OF SERVICE
# Highlights from Annual Reports
YEAR: 1983— 1984

## President's Message

Fiscal year 1983/84 was a challenging year for the Golden Gate Bridge, highway and Transportation District. Significant accomplishments were achieved when the District:

- Implemented measures to improve safety for motorists.
- Continued negotiations to acquire Northwestern Pacific Railroad's right-of-way.
- Established criteria for the design, development and testing of a moveable media barrier on the Bridge.
- Joined with other agencies and jurisdictions to mutually develop solutions to Highway 101 congestion;

Maintenance of the Bridge, support for solutions to congestion, and safety for those who use the Bridge guided the actions of the District.

**45 MPH Speed Limit.** A 45 MPH speed limit on the Golden Gate Bridge was implemented October 1, 1983, supported by a public awareness program and increased surveillance and enforcement by the California Highway Patrol. These actions were recommended by the Traffic and Safety Improvement Task Force created by the Board to study and evaluate traffic condition on the span.

**Movable Median Barrier Feasibility Study.** The engineering firm of Sverdrup & Parcel and Associates, Inc. was retained to determine the feasibility of installing a movable median barrier on the Golden Gate Bridge. Four movable media barrier concepts were evaluated. A barrier developed by Quick-Steel, Ltd. of Australia was considered the most appropriate barrier for the Bridge. However, Sverdrup & Parcel recommended that no barrier be installed as they considered the marginal traffic safety that might be provided by a movable median barrier would not warrant the estimated $12 million required to complete research and development, design, fabrication and installation of such a barrier.

In an effort to explore all possibilities, the Board encouraged others to submit designs and offered $50,000 for design and development of a movable median barrier that will meet all necessary operational and maintenance parameters at a lower cost.

**Northwestern Pacific Railroad Right-of-Way.** The District has acquired one mile of the abandoned right-of-way. During 1983-84, the District continued negotiations with Southern Pacific Company to acquire additional portions from Paradise Drive in Corte Madera to Highway 37 in Novato.

**San Rafael Transportation Center.** Work progressed on the Center. Planning studies and preliminary design plans will integrate the multi-modal transportation with any future use of the Northwestern Pacific Railroad right-of-way by transit rail or bus.

**Local Marin County Bus Service.** In July 1983, the Bridge District assumed operation of local bus service within Marin County at the request of the Marin County Transit District. The District is working with Marin County to seek solutions to the continuing problem of funding Marin bus service.

*In the rest of the World . . . .*
## HISTORICAL SNAPSHOT

**Events.** *U.S. economy expands; Big 3 auto makers sell 14-million vehicles. Daily volume of New York Stock Exchange exceeds 200-million shares for first time. Los Angeles replaces Chicago as U.S. second-largest city. Reagan reelected president.*

**Music.** *Prince, Purple Rain. Bruce Springsteen, Born in the USA.*

**Sports.** *Summer Olympics held in Los Angeles. Growing interest in fitness results in $10-billion sales of home exercise equipment. After Olympics, Peter Ueberroth becomes Commissioner of baseball.*

**Everyday Life.** *75 million in U.S. watch "The Day After," about nuclear attack. Home video-taping ruled legal by Supreme Court. The Yuppie Handbook, defines '80's icons. Donald Duck is 50.*

# 1984

**Highway 101 Corridor Plan Action Committee.** Traffic congestion problems and future transportation needs which affect Sonoma, Marin and San Francisco Counties are being investigated and analyzed by a 23-member Committee, composed of elected officials and representatives of cities, counties and agencies in the three counties. (They) will develop a coordinated, corridor-wide plan through the year 2005.

## Report of the General Manager

**Bridge Traffic.** Lower gas prices and lessening public emphasis on energy conservation have contributed to a decrease in the use of the District's transit services, resulting in record levels of traffic on the Golden Gate Bridge.

A total of 38,519,280 vehicles crossed the Bridge during fiscal 1983-84. On August 11, 1983, 119,629 vehicles traveled across the Bridge, the highest single day total since the Bridge was opened.

As of June 30, 1984, 976,520,840 vehicles have crossed the Bridge during its 47-year history.

**Bridge Closure Due to High Winds.** On Saturday, December 3, 1983, the Bridge was closed at 11:20 AM due to severe high winds. To lanes in each direction were opened to traffic at 2:55 PM. The Toll Plaza anemometer registered winds as high as 77.1 mph. The Bridge sustained no damage. This is the third time in its history that the Bridge had been closed to traffic in order to protect motorists during windstorms.

**Bridge Maintenance.** At 1:22 AM, on the night of November 4, 1983, the first of 747 modular deck units was lifted into place on the Golden Gate Bridge for the deck replacement. By June 30, 1984, 270 deck sections had been replaced and the $70.6 million was on schedule.

During the Deck Replacement Project, the shape of the roadway has changed due to the construction sequence of replacing the concrete deck with a light weight orthotropic steel deck. Flexibility is an inherent characteristic of suspension bridges and the present deflection of the structure is well within the limited designed by Chief Engineer Joseph Strauss. The roadway will return to its original profile as the project proceeds toward completion.

**New Electronic Fare Boxes.** The installation of 264 state-of-the-art fare boxes on Golden Gate Transit buses was completed in May 1984. The new fareboxes, designed to accept dollar bills as well as coins, are expected to provide more efficient fare collection and revenue processing.

**New Coaches.** The first of 51 new suburban-type buses began operating between Santa Rosa and San Francisco in September 1983. The new coaches, equipped with high-back seats, overhead luggage racks and reading lights, are the type favored by most Golden Gate Transit patrons. The new coaches are accessible and equipped to accommodate one or two wheelchairs.

**Methanol Fuel Demonstration Program.** The California Energy Commission selected Golden Gate Transit as the agency to test the use of methanol as an alternative fuel for internal combustion engines. The GMC and M.A.N. coaches are the first methanol-powered buses to be operated in regular service by a public transit agency in the United States.

**Diesel Conversion of G.T. San Francisco.** On December 2, 1983, the G.T. San Francisco was towed from the Larkspur Ferry Terminal and arrived in San Diego two days later. The vessel is the first of the three Larkspur ferries to be converted from gas turbine-waterjet propulsion to diesel twin-screw propulsion engines. The San Francisco will return to service in the fall of 1984, with the remaining two Larkspur vessels to be converted to diesel power shortly thereafter.

**Ridesharing Employer Outreach Program.** In March 1984, Ridesharing formed its 300th Vanpool. As of that date, about 2,000 commuters were traveling to and from work in Vanpool groups that received direct assistance from the Ridesharing Division. ☐

SIXTY YEARS OF SERVICE

# Highlights from Annual Reports

YEAR:1984— 1985

## President's Message

The past year signaled a renaissance of enthusiasm, unity of purpose and recognition of the interdependence between the Golden Gate Bridge, Highway and Transportation District and the Bay Area region.

**Marin County Local Bus Service.** Over the last ten years, transbay transit fares and gasoline prices have increased at almost the same rate. By 1984, local Marin fares exceeded the cost of operating a compact car and became detrimental to transit patronage and the District's goal of maintaining mobility along the Golden Gate corridor. After climbing to a high of 54% of all local Marin patronage in 1978, student ridership tumbled to 32% in 1984.

To turn around a deteriorating situation, the District, together with the County of Marin, took unprecedented actions. Elimination of Marin fare zones to create a uniform student fare and reinstatement of a student discount fare were implemented beginning July 8, 1985.

**Northwestern Pacific Railroad Right-of-Way.** The District acquired the (right-of-way) from Corte Madera Creek south to Paradise Drive, together with a half-acre parcel adjacent to the right-of-way, on December 28, 1984. In 1983, a 1.5 mile section, from Corte Madera Creek north to Bellam Boulevard in San Rafael was purchased.

**Traffic and Safety Improvements.** The Traffic and Safety Improvements Task Force was created in April 1983 to study and evaluate traffic mobility, safety, and traffic improvement project. Short-term projects that were identified — speed limit regulations on the Bridge and its approaches, increased enforcement by the California Highway Patrol, and installation of variable message signs on both end of theBridge — have been implemented.

The traffic safety study of the movable median barrier concept is being performed by the Traffic Institute, Northwestern University in Illinois. Twelve concepts of movable median barriers were accepted for the District's Movable Median Barrier Research and Development Program and held in abeyance pending completion of the Traffic Institute's report.

The Task Force's long-term project came to a close when further feasibility studies of installing a second deck on the Golden Gate Bridge were tabled. Its charge completed, the Task Force was disbanded in November 1984.

**Golden Gate Bridge District Museum.** The Board of Directors, on November 30, 1984, approved the formation of a non-profit "Friends of the Golden Gate Bridge" foundation and authorized the Attorney to files Articles of Incorporation and apply for tax-exempt non-profit status.

In 1977, the Board of Directors endorsed the concept of a Golden Gate Bridge Museum. Since then, $146,715 in proceeds from various revenue sources have been set aside in a Museum District Reserve Fund.

---

*In the rest of the World . . . .*

## HISTORICAL SNAPSHOT

**Events.** *Federal deficit hits $211.9 Trillion. U.S. dollar devalued around world. Wreck of the Titanic discovered by French explorers. A ban on leaded gasoline ordered by Environmental Protection Agency. A complete ban on tobacco advertising urged by AMA. Dow Jones sets record at 1553.17.*

**Films.** *Australian Rupert Murdoch buys 50% of 20th Century Fox. Back to the Future, Out of Africa, Ran.*

**Sports.** *49ers defeat Dolphins, 38-16, in Superbowl XIX. Nolan Ryan becomes first pitcher to strike out 4000 batters. Boris Becker, 17, becomes youngest and first unseeded player to win Wimbledon.*

**Everyday Life.** *Coke tries to change formula, relents, returns it as "Classic."*

# 1985

*[In addition to reports from the President and the GM, this year included another report — on the completion of the new roadway on the Golden Gate Bridge. In no-nonsense style, the story of the historic replacement, especially the last few hours, gained as much drama as the story of the Bridge's last rivet, but as you'll see, involved considerably higher technology.]*

## General Manager's Report

District changes in commute travel patterns have occurred within the Golden Gate Corridor. Recent studies revealed a majority of commuters now work in their home county.

Extensive office and retail developments in central and northern Marin County and rapid population grown in Sonoma County have redefined the demographics of the North Bay, altered commute origin and destination patterns, and caused severe traffic congestion between Marin and southern Sonoma counties.

By monitoring shifts in commute and off-peak travel patterns, District staff has anticipated and responded aggressively to changes created by construction activity. Transit routes and schedules are adjusted to ease congestion problems and to meet the changing travel needs of residents served by the District.

**Bridge Traffic Milestones.** During the month of February 1985, the District paid tribute to the one-billionth vehicle to cross the Golden Gate Bridge since opening day on May 28, 1937. A billion cars, an awesome figure, would form a line stretching around the earth 113 times or traveling to the moon and back five and a half times.

The Golden Gate Bridge carried a total of 38,630,790 vehicles during the fiscal year ending June 30, 1985. As Bridge traffic approaches 40 million vehicles annually, the standard for maximum practical capacity, estimated at 25 million a year, will have been exceeded by 60%.

**Diesel Conversion of Larkspur Ferries.** The First Larkspur Ferry to be converted to diesel from gas turbine-powered engines returned from San Diego at 4:00 AM on December 16, 1984. As the M.S. [instead of G.T. for "gas turbine"] San Francisco, it was placed back into service on December 24.

The GT Marin left for San Diego on January 2, 1985, arrived back in San Francisco on May 26, and reentered service June 3. The third diesel-converted vessel is anticipated to go into service on November 18, 1985.

For the first time in many years, midday and weekend Larkspur Ferry service was continued through December 31, 1984 and resumed May 1, 1985. The more fuel-efficient diesel engines will enable the District to provide a more comprehensive ferry schedule than was previously possible.

**Ridesharing Vanpool Program.** In June of 1985, the Board of Directors approved the reorganization of the Ridesharing Division. The District's Vanpool transition program will be turned over to RIDES for Bay Area Commuters, Inc., in the fall of 1985.

**District Engineer's Report.** On August 15, 1985, at 2:30 AM, exactly 401 nights and days of work after the first section of the 47-year-old reinforced concrete deck was sawcut from the Golden Gate and replaced with a modern orthotropic steel deck section, the 747th deck section was lowered into place, complete the structural work on the $65 million project.

This project replaced the entire original roadway slab and its supporting steel stringers.

Since the completion of the Golden Gate Bridge in 1937, the salt atmosphere had taken its toll. In 1976 it was determined that the chloride content of the original reinforced concrete roadway slab exceeded the threshold limits for rehabilitation and the roadway would have to be replaced in its entirety.

The original reinforced concrete sidewalks and their structural steel supporting system were in need of rehabilitation as well. This work was incorporated into the Bridge deck replacement contract.

All work involving actual deck replacement was performed at night. Low night-time traffic volume allowed the four lanes to remain out of service until 5:30 AM, at which time the Bridge was configured to accommodate the AM commute traffic.

Since the orthotropic steel replacement roadway element is approximately 40% lighter than the original reinforced concrete roadway element and its supporting stringers, consideration had to be given to achieving a balanced reduction in weight. Computation proved, however, that no part of the structure would be stressed beyond allowable unit stresses under combined loading condition, so deck replacement began at the north end of the structure and progressed uniformly across the Bridge to the south end.

The final 2" of epoxy asphalt concrete will be applied over the temporary riding surface during the Spring of 1986. □

## SIXTY YEARS OF SERVICE

# Highlights from Annual Reports

## YEAR: 1985— 1986

### President's Message

As we view the major events of the past year, the Golden Gate Bridge, Highway, and Transportation District is anticipating the Golden Anniversary of the opening of the Golden Gate Bridge to traffic in 1937. While preparing to celebrate its past, the District is planning for its future. These fiscal year 1985/86 programs, designed and implemented to carry the District into the next century, included:

replacement of the original Golden Gate Bridge roadway and sidewalks and contracting for its final paving;

purchase of half of the land for the San Rafael Transportation Center;

pursuit of Federal and State monies to fully acquire the Northwestern Pacific Railroad Right-of-Way;

expansion of Larkspur Ferry services and initiation of ferry shuttle bus service in San Francisco;

breaking of ground for the new Administration Building in San Rafael;

purchase of 21 new intercity coaches and four new short buses.

**Highway 101 Corridor Plan & Northwestern Pacific Railroad Right-of-Way.** The primary mission of the Golden Gate Bridge District is to maintain and operate the Golden Gate Bridge, ensure reasonable mobility across the Bridge, and, insofar as resources permit, provide public transit service in the Golden Gate Corridor from Sonoma and Marin Counties to San Francisco.

Travel patterns have changed considerably and congestion on Highway 101 has increased dramatically in the 1980's. In 1984, a Highway 101 Action Committee began an analysis of traffic congestion along the three-county corridor.

The first phase of the "Highway 101 Corridor Plan" offered three solutions to Highway 101 congestion, each of which called for acquisition of the Northwestern Pacific Railroad for use as an exclusive transit way.

Therefore, the Golden Gate Bridge District's number one priority is to acquire the NWPRR Right-of-Way and eventually construct a public transit guideway from Sonoma County to the Larkspur Ferry Terminal.

**San Rafael Transportation Center.** The District's priority capital project for the next five years is construction of the San Rafael Transportation Center. The Center will provide shelter, restrooms, a snack shop, and a clean well-lighted passenger waiting area, as well as improve transit accessibility for elderly and handicapped travellers.

---

*In the rest of the World . . . .*

# HISTORICAL SNAPSHOT

**Events.** *Space shuttle, Challenger, explodes 74 seconds after take-off. The U.S. national debt has doubled in five years. The Dow Jones average closed at 1895.95 for the year. "Hands Across America," to aid the homeless, involved 6-million Americans.*

**Science.** *Jeana Yeager, Dick Rutan fly around world without refueling. First laptop computer introduced.*

**Music.** *Phantom of the Opera opens in London.*

**Films.** *Blue Velvet, Jean de Florette, Top Gun.*

**Sports.** *Three-point goal adopted by NCAA. Greg LeMond becomes first American to win Tour De France.*

**Everyday Life.** *Statue of Liberty re-opened after three-year restoration, by Presidents Reagan and Mitterrand. Oreo cookie marks 75th anniversary. Popsickles change from two sticks to one stick.*

# 1986

### General Manager's Report

In 1986, falling gasoline prices and federally-mandated higher-mileage automobiles have been major factors in a decrease in commute bus patronage and Ridesharing and an increase in auto commute traffic. In addition, commute travel patterns have changed as major employers decentralize operations and move large numbers of workers to suburban locations in the East and North Bay Counties.

Reshaping the commute habits and attitude of the single occupant automobile driver represents one of the greatest challenges we face today.

**Bridge Traffic.** For the first time in the 49-year history of the Golden Gate Bridge, annual traffic across the Bridge exceeded 40 million vehicles.

During the past twelve months, 40,489,422 vehicles were recorded crossing the bridge. This volume surpasses the theoretical capacity by 60%.

A one-day traffic record was set on June 19, 1986, as 125,628 vehicles crossed the span in a 24-hour period. The record was broken the following day when the count reached 126,828. An average of 110,930 vehicles crossed the Bridge each day in FY 1986.

**Bus Replacement Program.** Eleven of the oldest General Motors Corporation buses were declared surplus and sold. Four new 30-foot buses were purchased from Gillig Corporation of Hayward, California. These new shorter buses will be used in West Marin and on local Marin routes.

**Larkspur Ferry Service.** The Golden Gate Ferry System is the most environmentally desirable transportation alternative to the private auto. The District continues to explore ways to make our ferry services more attractive to commuters.

The Larkspur Ferry services were expanded to a three-vessel operation on November 18, 1985, after the arrival of the third Larkspur Ferry to have its engines converted to diesel power.

By June, commuter service patronage was up 33% while overall ridership increased to 44%. In addition, cost reductions anticipated by the diesel conversion have met and exceeded expectations.

A new glass-enclosed passenger waiting area within the Larkspur Terminal was completed during 1986. This added passenger amenity provides weather protection for 126 passengers and increases the appeal of the ferry service to commuters during inclement weather.

**Sausalito Ferry's Fifteenth Anniversary.** August 15, 1985 marked the fifteenth anniversary of the inauguration of the Golden Gate Sausalito Ferry Service. Since service began, the M.V. Golden Gate has carried over 13.1 million passengers across the Bay between San Francisco and Sausalito. □

**Three Landmarks.** *In the days since the Golden Gate Bridge was built, San Francisco acquired two other significant landmarks. Both were landlocked and are visible in this photo. First, (center, right is Sutro Tower, tallest structure in the city, home of transmission sites for all of the major television stations and several FM radio stations. The other, less visible, but centered between the fourth set of suspender cables from the left, is the TransAmerica Pyramid, perhaps the most familiar office building in the world.*

## SIXTY YEARS OF SERVICE

# Highlights from Annual Reports

## YEAR: 1986— 1987

50th ANNIVERSARY! [Appropriately, this was the first four-color production of an Annual Report. The reason made spectacular sense, as illustrated by the colorful night photograph of the illuminated towers of the Bridge. Between covers, there are spectacular color photos of Bridgewalk 1987 showing a shoulder-to-shoulder crowd packing the roadway, as well as a center-spread night shot of the "waterfall" of fireworks reaching from the Bridge roadway to the inky waters of the Bay below. If ever there was a keepsake edition of an annual report, it is very likely to be this one.]

### President's Message

**T**he Dream Plus Fifty For many decades, it had been man's dream to bridge the Golden Gate. It has now been fifty years since the realization of that dream. Today the Golden Gate Bridge proudly stands as a world-famous monument to Chief Engineer Joseph B. Strauss and the energetic, determined, and courageous people who helped build it.

Since the Golden Gate Bridge opened for traffic on May 28, 1937, it has been a breathtaking showcase of architectural integrity, a shining symbol of human spirit, a memorable tribute to its designers and builders and a constant source of pride to those who operate and maintain it.

**The Celebration Day.** The excitement and thrill which accompanied the opening of the Bridge in 1937 was repeated and amplified on May 24, 1987, a day that created new lifelong memories.

"Pedestrian Day '37" became "Bridgewalk '87." A day when a great outpouring of people came to pay homage to the Bridge, and be part of a historical event.

The happy crowd included foreign visitors as well as individuals who came to relive their 1937 experience. Perhaps the most indelible image was the goodwill displayed by the throng of people who surged onto the Bridge during the start of the days celebration.

After the Bridgewalk, a parade of classic antique cars recreated the 1937 Opening Day Motorcade. Following the parade, the Bridge was opened to regular traffic. As a token of appreciation to the thousands of motorists who use the Bridge each day, the Board suspended collection of tolls for the day.

Afternoon and evening festivities continued on Marina Green and at Crissy Field, scene of a star-studded evening concert. The celebration included a stunning fireworks display which featured a brilliant pyrotechnic "waterfall" that showered from the Bridge roadway to the Bay below.

The original team of men who designed and build the Golden Gate Bridge envisioned that the Bridge towers would be illuminated at night. However, financial constraints of the Depression era eliminated the lighting plan during Bridge construction.

The designers' vision was not forgotten. Their dream became reality. At the climax of the evening's fireworks display ceremonies, newly-installed tower lights were switched on bathing the towers in light. This was truly a fitting and spectacular finale to an unforgettable day. □

*In the rest of the World . . . .*

### HISTORICAL SNAPSHOT

**Events.** *In worst stock market crash in U.S. history, the Dow fell 508 points on October 19. Close at year-end: 1938.83. Earthquake registering 6.1 shakes Los Angeles. Drug AZT approved by FDA. First ad for condoms to prevent AIDS plays on KRON-TV, San Francisco.*

**Science.** *Fibre optic cable laid across Atlantic. Digital audio tapes (DAT) go on sale.*

**Sports.** *Dennis O'Conner regains America's Cup yacht trophy after losing it to Australia. Little League celebrates 40th anniversary.*

**Everyday Life.** *More Americans watch football than baseball on television. Thirty states allocate funds to fight AIDS. The mini-skirt made a brief comeback.*

# 1987

## A Golden Anniversary Year

**Accolades and Artifacts.** The Golden Anniversary inspired tributes, symposia, civic galas, congratulations and warm wishes from historical, cultural, governmental, educational, business and labor organizations worldwide. Artists' and musicians' compositions acclaimed the past history, the present joy and the future hopes for the Bridge.

**Roundhouse Renovation.** The Roundhouse, an integral part of the Toll Plaza environment, was renovated and remodeled into an attractive Visitors' Information and Gift Center.

**Commemorative Gardens.** The promontory, north of the new Gift Center was transformed into landscaped gardens to commemorate the eleven men who perished during the construction of the Bridge.

**Northwestern Pacific Railroad Right-Of-Way Acquisition.** Consultants' studies of the right-of-way revealed grades and curves of the NWPRR were adequate to permit high speed modes of transportation, highway tunnels provide some grade separation, and many existing structures are suitable for limited adaptation, to transportation use.

A contiguous three-mile section south of Bellam Boulevard in San Rafael has been acquired. Negotiations continue with the S.P. Railroad for the ten miles from Bellam to Highway 37 in Novato.

**San Rafael Transportation Center.** The Center is the District's primary capital project for the next five years. It will consolidate on-street transfer stops of five transit operators into a common terminal, provide needed passenger amenities and improve accessibility for elderly and handicapped riders.

**Contra Flow Lanes.** Caltrans' studies of the evening traffic flow over Waldo Grade documented a substantial increase in southbound traffic, and that the special Waldo Grade "contra-flow" lane, used by Golden Gate buses to bypass northbound commute traffic, was not providing a significant advantage. Therefore, the contra-flow lane was discontinued in 1986.

## General Manager's Report

**Bridge Traffic.** Vehicle traffic across the Golden Gate Bridge totaled 42,220,298 for the year. A new all-time record occurred on June 18, 1987 when 132,674 vehicle crossed the Bridge in a twenty-four hour period, nearly 6,000 more than the previous record.

By the end of FY 1986/87, 1,095,861,350 vehicles had traveled across the Bridge since it opened.

**Bridge Deck Replacement Project.** The final phase of the Bridge Deck Replacement Program was completed on December 22, 1986, when the final two-inch thick surfacing of epoxy asphalt had been applied to the new orthotropic deck. Six thousand tons of epoxy asphalt was applied in just twenty nights of work.

Replacing the original reinforced concrete deck and its supporting system with a modern lightweight steel plate deck reduced the weight of the Bridge by 11,500 tons.

**Bridge Rehabilitation Programs.** In 1985 restoration began on both the San Francisco and Marine approach spans to the bridge. The restoration, undertaken by District maintenance workers, is expected to cost $6 million.

The pier and fender system of the San Francisco tower was inspected and found structurally sound; however, recommendations were made to recondition and strengthen the fender.

**Larkspur Ferry Patronage Increases.** Patronage on the Larkspur ferries increased 17% to 846,378 passengers. A reduction in operational costs was realized when the vessels' gas turbines were replaced with diesel engines. During the fiscal year, more than 1.4 million passengers crossed San Francisco Bay on Golden Gate Larkspur and Ferry services.

**Bus Replacement Program.** Twenty-seven of the original fleet of General Motors buses were retired from service and sold. The sale price represented 35% of the original purchase price. Golden Gate Transit's enviable reputation for outstanding maintenance ensures a ready market for the District's surplus rolling stock.

**San Rafael Facilities Expansion.** A new District Administration building was completed in March. Several District administration departments, formerly housed at various locations, immediately occupied the new two-story structure. Design and engineering work to remodel the existing bus administration building commenced in March. The interior remodeling is expected to take seven months.

**Bus Service.** Beginning in 1982, Golden Gate Transit's bus patronage followed a national downward trend. In July 1986, commute bus service was reduced by 9% to coincide with demand for service and to reduce operating expenses. Seat occupancy averaged 63%, considerably less than the District's standard of 80% occupancy.

**District Employees.** *[Each year, the Board, as represented by the President, and the Administration and Operations, as represented by the General Manager, have graciously acknowledged and thanked the employees of the District for their continued loyalty and dedication. We have not included these comments in these edited selections, but the 50th Anniversary message to the workers reflects the spirit and sincerity of all the years.]*

This hallmark year has been one of celebrating the past. I wish to express my gratitude to each and every employee, whose support and hard work helped bring great credit to the Golden Gate Bridge District. We can take great pride in the knowledge that, while the eyes of the world were focused on our great Bridge, we continued to carry out our prime mission of serving the daily needs of the public. We can now look forward to the future years of providing continued high quality service to San Francisco and the Golden Gate Corridor. □

**Here They Come!** *The anticipation and excitement surrounding Pedestrian Day in 1937 was repeated and amplified on what has become known as "Bridgewalk 1987" - an event to commemorate the 50th Anniversary of the opening of the Bridge. As happened five decades earlier, crowds began to gather early (the street lights are still on) and jostle toward the Bridge. Here celebrants crowd onto the span which had been closed to traffic for the occasion.*

**And They're Still Coming!** *Hours later the great horde of well-wishers continued to push onto the Bridge. Note the swell of the crowd pressing through the Toll Plaza area. At its peak, the estimate of "Bridgewalk 1987" participants was placed at more than 300,000. As a result, the Bridge roadway, which has a graceful natural arc between the towers was flattened. Bridge* *engineers explained that no harm resulted and when the Golden Gate Bridge was reopened to traffic a few hours later, it returned to its natural position. Bridge lovers were quick to explain the structure had survived quakes and wind, it certainly could survive a celebration of fifty years of unexcelled service to Northern California.*

SIXTY YEARS OF SERVICE

# Highlights from Annual Reports

YEAR: 1987— 1988

## President's Message

**The Challenge of the Golden Gate Bridge.** Throughout the 60 years of its existence, the Golden Gate Bridge has faced and solved seemingly impossible problems. First and foremost, the Golden Gate Bridge was built, despite lack of faith by man, because of the vision of individuals — such as Joseph Strauss and A.P. Giannini — and of the elected representatives of the six California counties represented by the Board of Directors.

**The Bridge, The Budget and Public Transit.** Then, in 1969, with the passage of AB 584 by the California legislature, the Bridge District was presented a new challenge: to create a system of buses and ferryboats to offset increased traffic in the Golden Gate Corridor and by this means to decrease congestion on the Golden Gate Bridge.

The greatest problem addressed by the Golden Gate Bridge in the past year was a long-projected budget shortfall for the Fiscal Year 1988-89 of about $5.8 million. Unlike other Bay Area public transit systems, the Golden Gate Bridge District has no dedicated revenues from sales or property taxes. It must primarily rely on surplus Bridge tolls and farebox revenue to operate its buses and ferryboats.

The District must have and does have a balanced budget. But as always this could only be achieved by increasing revenues, decreasing expenses, or both.

The Board of Directors of the Golden Gate Bridge District thoroughly and realistically evaluated all options and explored all avenues of cost savings. Yet, the District also needs to continue our commitment to the needs of the North Bay commuter for reduced congestion on the Golden Gate Bridge, an object which is achievable only by continuing our commitment to public transit. Thus we have moved to raise tolls from their present one dollar five days per week and two dollars two days a week, to a regular discount toll of $1.25 per day for a six-month period, starting in January 1989, and $1.67 thereafter. The toll for a casual crossing of the Bridge (without pre-purchased discount tickets) will at that time be two dollars. The complex problems involved in funding our transit system require us to work together regionally and to listen to the needs of each segment of the public, which the District Board of Directors did through numerous public hearings throughout our service area. Virtually all the public officials in the region and many members of the public urged the District not to further reduce transit service (as was done in 1987), and also not to increase transit fares as a means of balancing the Budget. Instead, they encouraged the District to reduce costs where possible without compromising transit quality, to reduce fares, if possible, and to increase revenues by increasing Bridge tolls.

**District Leads on Accessible Transit.** The Golden Gate Bridge District reached a milestone this year in its continuing leadership role in making public transit fully available to persons with disabilities. Members of the disabled community lauded the District's policy, calling it the most progressive of any public transit agency in the United States.

---

*In the rest of the World . . . .*

## HISTORICAL SNAPSHOT

**Events.** *Economy grows, inflation increases. Iran-Contra figures (Gen. Poindexter, Lt.Col. North) indicted. Worst forest fire season in U.S. since 1919. Bush and Dukakis nominated for president.*

**Science.** *Stealth bomber goes on public display. Stephen Hawking publishes, A Brief History of Time.*

**Books.** *Thomas Harris, The Silence of the Lambs. Kobo Abe, The Ark Sakura. Salman Rushdie, The Satanic Verses.*

**Sports.** *Steffi Graf becomes third woman to win Grand Slam, four major tennis tournaments; she also wins an Olympic gold, as added to Olympic Sports. Olympics held in Seoul, Korea.*

**Everyday Life.** *45 million Americans now have cable-tv service. The Bolshoi Mak (Big Mac) goes on sale in Moscow. Fax machines are booming.*

## 1988

**Sister Bridge Affiliation.** In the spirit of Pacific Rim cooperation and friendship, the Golden Gate Bridge District and Japanese authorities signed a Sister Bridge Affiliation to honor the completion of Japan's Seto Ohashi "Grand Bridge."

A delegation of District and Bay Area officials participated —at their own expense — in the opening ceremonies of the Seto Ohashi Bridge in April of 1988. The span over the Japanese Inland Sea is 8 miles long and cost $8.5 billion to complete after 20 years of planning and design, and 9 years of construction. The toll is the equivalent of $45 U.S. — one way — for automobiles, while most people cross on the railroad trains which zip back and forth on the lower deck of the span.

## General Manager's Report

**Bridge Traffic Increasing.** A new, all-time traffic record was set on August 7, 1987, when 133,206 vehicles crossed the Golden Gate Bridge during a 24- hour period. 43,457,470 vehicles cross the bridge (this) year.  By June 30, 1988, a total of 1,142,318,820 vehicles had cross the Bridge since its opening in 1937.

**Bus Transit Ridership Declines.** Total bus patronage for this year was 7,332,141, a 0.9 percent decrease from the previous years total. The Bus Transit Division continued to make adjustments in service to maintain an efficient ratio of seats per passengers and to control costs.

District buses have carried a total of 116 million passengers since inception of service in 1971, which equates to a reduction of about 90 million autos on Highway 101 and the streets of San Francisco.

**Ferry Transit Ridership Growing.** Overall, passengers riding the Larkspur ferries were up 5 percent this year and commuter ridership was up a remarkable 12 percent. A total of 885,886 passengers traveled by the Larkspur ferries.

Sausalito ferry ridership decreased by 4 percent, totaling 566,196 passengers for the year.

A new record was set September 17, 1987 when Pope John Paul II visited San Francisco and a total of 9,746 passengers were carried across the Bay by ferry.

**Bridge Safety Improvement.** During the year, the California Highway Patrol (CHP) conducted radar surveys on the Golden Gate Bridge which showed that 85 percent of the motorists crossing the Bridge were violating the established 45 mile-per-hour speed limit, and travelling at speeds as high as 55 to 59 miles-per-hour. Following several fatal crashes on Doyle Drive and the Bridge in the Spring, the CHP assigned a special enforcement squad of six motorcycle officers to patrol the Bridge through June, 1989.

A five-fold increase in traffic citations at the beginning of the safety campaign leveled to about three-fold in June, 1988. The immediate result of this special enforcement was a marked decrease in accidents on the Bridge.

**Transit Stoppage Congests Bridge Span.** Bus drivers, members of the Amalgamated Transit Union, Local Division 1575, unable to agree to terms with the District on a new Memorandum of Understanding, stopped work between January 23 and 31, 1988. As a result, both bus transit and ferry transit services were interrupted.

Due to the temporary discontinuation of transit service, vehicles crossed the Bridge in record numbers during the 6-10 AM commute period. The highest single period total was 25,074.

**Ridesharing Program Honored.** The District received the Metropolitan Transportation Commission's Grand Award for 1987 for promoting Ridesharing.

To provide more efficient Ridesharing assistance to North Bay counties, the District supplemented its efforts in this activity by contracting with RIDES for Bay Area Commuters, Inc., with the intent of transferring the District's Ridesharing program to (them) in 1989.

**Bus Transit Division Projects.** This year 21 new buses, each equipped with wheelchair lifts, were acquired from Motor Coach Industries at a total cost of $4.4 million. The District plans to replace 30 of its older buses each year. There are currently 239 buses in the District fleet. By agreement with the California Energy Commission, the District continued to participate in the methanol-power bus demonstration project.

Ferry feeder bus service was extended to the San Marin area in Novato, and the feeder bus run south of Market Street in San Francisco was extended to the Civic Center. □

# SIXTY YEARS OF SERVICE

# Highlights from Annual Reports

## YEAR: 1988— 1989

## President's Message

The Golden Gate Bridge, Highway and Transportation District continues to be a dynamic organization vigorously meeting the challenges given it by the California State Legislature 20 years ago.

During the summer of 1988, the Board of Directors adopted a Toll, Fare and Service Improvement Plan which called for a Golden Gate Bridge toll increase to $2.00 and a program of transit service improvements and fare reductions. On January 2, 1989, the new Bridge tolls and reduced transit fares went into effect.

**Reasons for Change.** The program of changes was adopted to accomplish a number of important District goals:

To solve a projected revenue shortfall estimated at more than $38 million over the next five years. Because the District has no sales or property tax authority, its operations are financed by Bridge tolls, transit farebox revenues and funds made available by local, state or federal sources. Since state and federal transit funding has been declining for some years and increases in transit fares tend to reduce ridership, Bridge toll revenues were the primary focus for addressing future District financial needs.

To improve and expand transit services, particularly for Transbay bus commuters.

To address a growing need for commute transit service between Sonoma and Marin Counties.

To stem the annual increase in vehicles crossing the Golden Gate Bridge into San Francisco.

To finance safety and other improvements to the Bridge and Toll Plaza, to fund a study on the feasibility of providing a second Bridge deck for transit use, and to finance a major inspection of the main Bridge cables — the first in 20 years.

**Effects on Congestion and Ridership.** Results of the toll and discount program are outstanding.

The first six months of operation for the program resulted in decreased commuter congestion on the Bridge, high levels of toll discount ticket use, and increased ridership on Golden Gate Transit bus and ferry services. The Directors voted to restore 11 Transbay runs that had been reduced in earlier years.

**Looking to the Future.** Over the next 20 years, the (101 Corridor Action Committee) plan envisions spending an estimated $1.3 billion to widen and improve Highway 101 through (Marin and Sonoma Counties), adding high-occupancy vehicle (HOV) lanes for buses and carpool vehicles, building a transit system to use the NWP right-of-way and increasing ferry service with new, high-speed vessels. A proposal to ask voters in Marin and Sonoma to approve adding one cent to the existing sales tax will be submitted to the voters of each county in the near future, with the new money earmarked for transportation and transit.

**Progress Toward Full Accessibility.** The District was included in a 1989 list of America's ten most accessible transit systems for its efforts in making Golden Gate Transit useable by disabled persons. The ranking was issued by Americans Disabled for Accessible Public Transportation (ADAPT).

[The cover for this year's Report was one of the most unusual in the history of the District. Inspired by the 50th anniversary of the Bridge, the San Francisco Quilters Guild hosted an international competition. One hundred and twenty-five quilters submitted their unique, colorful blocks. The best twenty were chosen to be assembled in The Bridge Quilt, which was presented to the District at a special ceremony.

The quilt, seven and one-half feet by eight and one-half feet depicts a variety of Bridge scenes, including the towers projecting through the fog, a cross section of the main cable, even a tribute to the "Halfway to Hell Gang" — workers who survived falls during the construction. The four-color photo was taken by Bob David, the Bridge photographer.]

---

*In the rest of the World . . . .*

## HISTORICAL SNAPSHOT

**Events.** *Students march on Beijing's Tiananmen Square. The Exxon Valdez runs aground in Alaska, spills 11 million gallons of oil. Emperor Hirohito dies; Crown Prince Akihito succeeds him. Loma Prieta earthquake hits Northern California. Bush and Gorbachev declare end of Cold War.*

**Books.** *Amy Tan, The Joy Luck Club. Kazuo Ishiguro, Remains of the Day.*

**Films.** *Batman, Driving Miss Daisy, When Harry Met Sally.*

**Everyday Life.** *President Bush declares his distaste for broccoli; growers protest, send free broccoli to the White House. Evangelist Jim Bakker sentenced to 45 years for fraud.*

# 1989

## General Manager's Report

After six months experience with the $2 toll increase, the results appear significant and positive. Vehicular traffic crossing the Golden Date Bridge was only 43,391,562, a decrease of 0.15 percent. About 70 percent of motorists crossing the Bridge during commute hours use the new toll tickets. This has resulted in faster, improved traffic flow through the Toll Plaza.

Golden Gate Bus patronage has increased approximately 13 percent since the new transit fare discount. Combined total ridership on the Larkspur and Sausalito ferries was 1.5 million, an increase of 2.9 percent over the previous year.

**Bridge Division.** Directors authorized an engineering study to evaluate the structural capability of the Bridge to support a second deck dedicated to fixed-guideway transit. The study will consider the structural safety of the Bridge as well as compare feasibility and costs of constructing a BART type rail system, a light rail system, a monorail, a railbus or any other rail system that may be identified as a practical alternative.

The California Highway Patrol assigned special units, as available, to improve drivers' observance of the 45 mph speed limit on the Bridge. The 12 months ending April 30, 1989, fatalities on the Bridge were reduced 75 percent from a year earlier, and the overall number of injury accidents was reduced 23 percent. The number of speeding citations rose 97 percent.

**Bus Transit Division.** In February 1989, the California Department of Motor Vehicles commended the Golden Gate Transit for its driver safety training program and District drivers for their excellent safety record.

The District awarded a contract for 80 new passenger lift-equipped buses to replace old buses, many of which have been in service since the early 1970's.

The Board formed a Golden Gate Transit Bus Passengers Advisory Committee representing bus riders from San Francisco, Marin, and Sonoma Counties. The Committee will review both existing and planned bus services.

**Ferry Transit Division.** .Previously, free feeder service in Marin was offered only during AM hours to the ferry terminals. Beginning July 1, 1989, feeder buses will also be free during evening hours. Free feeder bus service to and from the San Francisco ferry terminal is also provided during commute periods.

**Capital Projects.** Work was completed on expansion and remodeling of the Bus Transit Administration building and expansion of the Bus Maintenance building in San Rafael. Work was completed to enlarge parking at the Larkspur ferry terminal, adding 261 stalls to the previous 996 spaces, as well as dredging of the Larkspur ferry channel.

The District Board approved revised plans for an integrated transportation center in San Rafael, designed to replace the Golden Gate Transit transfer point at Fourth and Heatherton Streets with a new convenient structure serving both Golden Gate Transit and other modes of public transportation as well as Greyhound Lines.

The District's Gift Center, located at the Toll Plaza, has become an increasingly popular spot for tourists and visitors. The Center's operating hours during the tourist season were extended and the sales staff was increased. Remodeling has provided more sales space. □

# SIXTY YEARS OF SERVICE
# Highlights from Annual Reports
## YEAR: 1989— 1990

*[In times past, the District surveyed the past and contemplated the future. However, a major natural disaster in 1989 broke this chain of study and, with it, created the need to examine extraordinary expectations. The event was fearful to contemplate and impossible to predict. It is now referred to as the Quake of '89. It occurred on October 17th at 5:04 p.m. — along the Loma Prieta Fault. The magnitude was 7.0 on the Richter Scale. The response of the people who staff and manage the Golden Gate Bridge, Highway and Transportation District was swift and sure.]*

## The Quake: President's Message

The earthquake of October 17, 1989 was a momentous event. The Bridge withstood the most savage temblor to strike the Bay Area since the great earthquake of 1906. The only significant damage to District property was to the fueling facilities at the Larkspur Ferry Terminal, which did not disrupt ferry service.

District staff responded to the emergency in excellent fashion, taking immediate action to alleviate the resulting myriad of traffic problems; thus making travel more bearable during this difficult time.

Bridge tolls and high occupancy vehicle lane rules were temporarily suspended. Extra buses were scheduled, and ferry runs were added to help smooth the commute for regular Highway 101 Corridor travelers and a temporary flood of 30,000 to 40,000 new daily travelers from East Bay counties.

On October 27, 1989, tolltakers recorded an all-time record of 164,414 vehicles crossing the Golden Gate Bridge in 24 hours.

Total weekday ferry ridership increased 60 percent during the months of October and November 1989 and transbay/weekday bus ridership increased 8 percent during the same period.

Japanese citizens used the Golden Gate Bridge/Grand Seto Bridge Sister Program to channel earthquake relief to northern California. At the Board of Director's meeting of December 22, 1989, (the Board President) accepted a check for $22,720 from residents of Okayama Prefecture. The money was delivered to the American Red Cross of the Bay Area.

The gift is yet another example of the goodwill created by the relationship between the District and the Japanese leaders responsible for the magnificent bridge across the Inland Sea.

## The Quake: General Manager's Message

District employees responded in an outstanding fashion to meet unexpected demands caused by the October 17, 1989 earthquake. Golden Gate Bus Transit drivers who were at Candlestick Park with busloads of passengers for the World Series game acted calmly and reassuringly, and transported their passengers safely back to Marin County despite power outage and extreme congestion in San Francisco.

---

### In the rest of the World . . . .
### HISTORICAL SNAPSHOT

*The Bridge was conceived and built during an important, expanding time in history. It's birth was difficult; it exists in a real world. These are a few of the events, near and far, that happened concurrently with its development and growth. As you'll see, from time to time events occurring far from San Francisco Bay held influence over the business of the Golden Gate Bridge.*

**Events.** *Americans with Disabilities Act gives rights of access to public facilities, jobs equality to disabled. Boris Yeltsin elected president of Russian Federation. Iraqui forces invade Kuwait; President Bush sends U.S. forces to Saudi Arabia. East and West Germany reunify. Nelson Mandela released after 27 years in prison.*

**Science.** *Space shuttle Discovery places Hubble Telescope in orbit. Japan places small satellite in lunar orbit. Pierre Chambon announces gene discovery important in development of breast cancer.*

**Films.** *Cinema Paradiso, Dances With Wolves, Dreams, Silence of the Lambs.*

**Deaths.** *Lewis Mumford, Greta Garbo, Alberto Moravia, Leonard Bernstein, Aaron Copland.*

**Sports.** *Martina Navratilova wins 9th Women's Singles at Wimbledon. Boxer Buster Douglas wind world Heavyweight title in Tokyo, knocking out Mike Tyson; later in year Douglas loses title to Evander Holyfield.*

**The Golden Gate Bridge** *begins an earnest search for funds to retrofit the Bridge as mandated by law following the '89 Loma Prieta earthquake.*

## 1990

Although the San Francisco Ferry Terminal sustained damage and was temporarily inaccessible, full loads of passengers were embarked from the District's alternate ferry facility located at Pier 1. All ferries continued to operate until every passenger travelling to Marin County was accommodated.

District Officers, Directors and Bridge staff temporarily suspended tolls and initiated other measures to minimize traffic backups on the Bridge. The District accepted toll tickets from State bridges in the Bay Area and made special arrangements with trucking fleets that normally did not use the Golden Gate Bridge.

Motorists who held unused Golden Gate Bridge toll discount tickets that had been scheduled to expire December 31, 1989 were allowed to use them for two extra months.

Following the earthquake, District employees asked bus and ferry riders to donate food and clothes for relief to the Watsonville and Santa Cruz areas in cooperation with the Marin Food Bank. Transit riders donated food and clothing. On their own time, District employees continued collecting relief supplies at supermarkets in Marin County, and participated in an airlift of the donated materials to Watsonville.

*[Responding to a major emergency was made to seem almost routine in the measured tones of the Annual Report. However, despite the fact the District did not choose to pat itself on the back, commuters, Bridge lovers and ordinary citizens saw how completely "Golden Gate people" pitched in to help in a time when cooperation and mutual aid was so important to the Bay Area Community at large. That accomplished, the District went back to work on its usual slate of problems, eyes firmly fixed, as always, on the needs of the future.]*

### President's Message

As we begin the final decade of the twentieth century, the Golden Gate Bridge, Highway and Transportation District is playing a partnership role in the improvement of transportation and transit service for the North Bay and Redwood Empire in the twenty first Century.

### Northwestern Pacific Railroad Right-of-Way.

On June 1, 1990, the District joined leaders of the Northwestern Pacific (NWP) Railroad Right-of-Way Task Force, the County of Marin, and the Marin County Transit District in signing a historic agreement with officials of the Southern Pacific Railroad. The pact establishes permanent public

ownership of the railroad right-of-way from Larkspur northward to Willits and eastward from Novato to Lombard in Napa County.

Total acquisition of the 151 mile right-of-way will be approximately $37.1 million.

Many problems remain before mass transit rolls on the right-of-way, but (the Board) is confident that this year will be remembered as a milestone in the history of the District's support for public mass transit.

### Transit Deck Study.

(The President) encouraged a transit deck study to determine whether the Golden Gate Bridge is capable of supporting a second deck for transit use by light or heavy rail vehicles, buses, or other forms of mass transportation.

The consultant selected to perform the study, T.Y. Lin International, concluded that a transit deck could be built within the structure below the existing roadway without significantly changing the appearance or structural integrity of the span.

With this question answered, the appropriate governmental agencies can address whether a transit deck should be added to improve transportation between San Francisco and counties north of the Bridge.

### Marin-Sonoma Commute Service.

District staff and employers in Marin and Sonoma counties have cooperated to design a new bus service for Sonoma residents who travel to job sites in Marin County.

The service is planned to begin in November 1990, with District commuter fare discounts similar to those provided for transbay commuters. The Board of Supervisors of Marin County has allocated funds to further subsidize fares on the new bus routes.

### Electronic Toll Collection Plans.

This year saw significant progress on proposals to smooth Bridge traffic through the Toll Plaza by the use of electronic toll collection. The General Manager organized an East Coast symposium on the state-of-the-art technology which was attended by interested officials from throughout the world.

This spurred interest by the California Department of Transportation and the California Legislature in creating an electronic toll collection system that could be installed uniformly on bridge and toll roads throughout the Bay Area and elsewhere in the State. The first field tests may be possible on the Golden Gate Bridge within 18 months.

## General Manager's Report

**Alternative Fuels.** — District Directors and staff devoted considerable effort to informing lawmakers, public transit and environmental agencies of District concerns about new clean air standards under discussion in Washington, D.C. The District urged that extensive tests of alternative fuels by large public transit fleets be made before the Federal Government adopts new clean air standards that could mean the end of diesel as the primary bus fuel. The District made wide distribution of a report which outlined its six years of experience testing methanol-based buses in a joint effort with the State of California.

The District's report indicated that abandoning the use of diesel fuel for public transit buses would harm the environment more than it would help.

A. Transit fares may have to increase by as much as 75 percent to balance $21 million in higher costs for fuel, new buses and maintenance facilities.

B. As a result, bus ridership would decrease by 35 to 40 percent, about 2.5 million patrons a year, undermining federal goals of persuading motorists to patronize mass transit.

C. Highway 101 would become even more crowded with single-occupant vehicles, about 2,050 cars during commute periods. This would increase congestion on the Bridge.

D. Golden Gate Corridor gasoline consumption would increase by 1.2 million gallons, adding to air pollution.

The District proposed that large-scale tests of alternative-fuel buses be conducted by at least four public transit systems in various regions of the United States before new federal emission standards take effect.

**Earth Day.** — The District participated vigorously in Earth Day observances on April 22, 1990, with displays in the cities of San Rafael and Santa Rosa. In addition, flyers promoting the use of transit and Ridesharing were handed out to motorists passing through the Golden Gate Bridge Toll Plaza for three days. The District furnished California Highway Patrol officers with complimentary bus and ferry tickets to pass out to motorists in an effort to encourage transit use and reduce congestion on Highway 101.

**Bridge Division.** — Total traffic was 43,900,850 vehicles. The October 17, 1989 Loma Prieta earthquake which closed the San Francisco-Oakland Bay Bridge for one month increased Golden Gate Bridge traffic by an estimated 226,000 vehicles.

**Seismic Study.** — T.Y. Lin International is conducting a study of the seismic stability of the Bridge and its approaches in light of new technical information resulting from the Loma Prieta earthquake and new standards for construction in areas of seismic activity.

**"Physical" for Span.** — (An engineering firm) was awarded a contract to perform a thorough inspection of the condition of the Bridge's main cables.

**Sister Bridge.** — A delegation of Japanese visitors attended the Board of Directors meeting on July 28, 1989, honoring the first anniversary of the Golden Gate Bridge Sister Bridge relationship with the Grand Seto Bridge in Japan.

**Traffic Safety.** — The District extended its contract with the California Highway Patrol to provide increased traffic enforcement on the Bridge and its approaches on Friday and Saturday nights through June 1991. The goal is to prevent traffic accidents caused by intoxicated drivers. There were not any fatality accidents on the Bridge and its approaches during 1989.

## Bus Transit Division

**Bus Patronage Gains.** — Overall bus ridership is up markedly. For the fiscal year 1989-90, ridership was 8.3 million.

**San Rafael Transportation Center.** — Engineering staff completed final plans and specifications for the future transit hub in the eastern area of downtown San Rafael to be build at an estimated cost of $3.9 million.

**Rider Participation.** — The Bus Passenger Advisory Committee completed its first year of regular meetings with District staff. The Committee, composed of Golden Gate bus riders, was established for a one-year period, holding its first meeting on July 12, 1989. In July 1990 both its members and District Directors agreed that the Committee should be made permanent so that it might continue to assist the District.

## Ferry Transit Division

**Ferry Patronage Gains.** — Combined total ridership on the Larkspur and Sausalito ferries was 1.6 million. The ten millionth passenger to use the Larkspur ferry was feted on September 18, 1989, and received a one-year free commute on the ferry. □

*The Honshu-Shikoku Bridge in Japan is the world's longest bridge. The Shimotsui-Seto suspension segment shown here is an official "Sister" of the Golden Gate Bridge.*

# The Golden Gate Bridge Has A SISTER BRIDGE

Even though it is fifty years younger, it is in fact a *big* sister in terms of size, construction cost and toll charges. It is Japan's Seto Grand Bridge, which joins that country's largest island, *Honshu*, with its smallest, *Shikoku*, in a very large bay called the Seto Inland Sea.

The arrangements whereby two great bridges figuratively joined hands across the world's largest ocean is a historically significant contribution to international cooperation and understanding. The collective goodwill of two groups who share interest and pride in their respective bridges *and more*, is reflected in the proposal from the Seto Grand Bridge representatives to the Bridge Directors. □

January 28, 1987

Mr. Gary Giacomini
President of the Board of Directors,
Golden Gate Bridge, Highway and
Transportation District

Dear Mr. Giacomini:

Please, first accept our most sincere congratulations on the completion by San Francisco's Golden Gate Bridge of 50 golden years. We know that the magnificent meetings, events and exposition planned to celebrate this memorable anniversary in May will be a very impressive success.

We very much hope that this letter may be a first step leading to the conclusion of a sister bridge agreement linking San Francisco's Golden Gate Bridge and Japan's Seto Grand Bridge (the world's longest two-tiered bridge and one of our country's greatest engineering achievements of this century).

The vast expanse of the Pacific Ocean lies between our two countries, the United States of America and Japan are very close next-door neighbors. Moreover, thanks to advances in science and technology, the significance of even such a vast obstacle is being drastically reduced, one might even say year by year.

We feel that this is a most auspicious time to offer our most friendly congratulations and propose the conclusion of a Sister Bridge Agreement. We would like also to propose the following:

1. Sending of a Japanese non-government mission to the Golden Gate Bridge on the occasion of the 50th Anniversary celebrations in May;

2. Exchange of commemorative monuments by San Francisco and Kagawa prefecture;

3. Inviting of 50 Americans to the Seto Grand Bridge on the occasion of the bridge's inauguration in the spring of 1988;

We plan to make a 3-day visit to San Francisco and the Golden Gate Bridge (February 17th-19th), and would greatly appreciate the opportunity of a preliminary meeting with you.

We greatly hope that you will welcome our proposal, and look forward to hearing from you.

Sincerely yours,

Takuya Hirai
Representative of the Sister
Bridge Preliminary Committee

Vice-Chairman
Seto Grand Bridge Expo '88

SIXTY YEARS OF SERVICE

# Highlights from Annual Reports

YEAR: 1990— 1991

## President's Message

Fiscal Year 1990-91 has been an eventful and significant time for the District.

The major effort for the Board this year and beyond is to secure funding for the seismic retrofit of the Bridge, while continuing to improve and expand transit service for the District's bus and ferry riders. The Golden Gate Bridge is the major artery connecting San Francisco with the counties to its north. The Bridge was designed and built in the 1930's before the advent of modern seismic engineering.

The 1989 Loma Prieta Earthquake caused severe damage to many structures in the Bay Area. The Golden Gate Bridge suffered no significant damage. However, the Governor's Board of Inquiry recommended retrofitting all transportation structures of regional importance. Even before the State's report, the District took action to employ consultants for a state-of-the-art seismic evaluation of the Bridge.

The consultant hired for the seismic study found that the Bridge may be vulnerable to damage and temporary closure if an earthquake with magnitude 7 or greater occurs near the Bridge. Earthquake protection work, estimated to cost about $128 million in 1991 dollars, was recommended by the Consultant. It is estimated that replacement cost of the Bridge would cost $750 million to $1 billion.

District Directors and staff are making concerted efforts to obtain federal and state funding assistance for the retrofit work.

Engineering for the retrofit is expected to cost approximately $10 million over the next two years, followed by three years of actual construction on the Bridge. The consultant is performing the environmental assessment and additional preliminary engineering. The District is preparing a request for proposals for final design work. In addition to the seismic retrofit work, the main span will be retrofitted to improve the stability of the Bridge in severe windstorms.

## Improving District's Financial Status

During the first half of fiscal year 1990-91, it became obvious that District revenues would not be sufficient to perform the seismic retrofit and improve transit services, or even maintain transit service at existing levels. Therefore, District Directors and staff extensively studied alternative courses of action, including increases in Bridge tolls and transit fares which are the only sources of income within the authority of the District. Unlike other transit agencies in California, the District has no tax collecting powers.

District staff began reviewing financial trends in December 1990. Their computer models estimated that under the old toll and fare structure, revenues would be approximately $102 million short of meeting District needs and proposed improvements through fiscal year 1995-96.

Contributing to the shortfall were the seismic retrofit program, higher operating costs for Golden Gate Transit, increased medical costs and other expenses.

---

*In the rest of the World . . . .*

### HISTORICAL SNAPSHOT

**Events.** *U.S.-led coalition launches Operation Desert Storm. Hundreds of Kuwait oil wells set on fire as Iraqi army flees. President nominates Clarence Thomas to Supreme Court. Edith Cresson becomes first prime minister of France. U.S. Forces in Gulf War include 30,000 women.*

**Books.** *Jean Rouaud, Les Champs d'Honneur. John Updike, Rabbit at Rest. Bret Easton Ellis, American Psycho.*

**Films.** *Beauty and the Beast, Terminator 2, Thelma and Louise.*

**Everyday Life.** *Seattle introduces two cultural phenomena, the coffee culture and grunge music.*

1991

After holding public hearings in San Francisco, Marin and Sonoma Counties, the Board of Directors voted to raise tolls on the Bridge to $3 cash for autos, effective July 1, 1991. This was a $1 increase from the $2 cash toll charged since January 2, 1989.

The Five-Year Financial Plan adopted May 31, 1991 by the Board of Directors includes not only the $3 cash toll, but also future changes in the Bridge toll ticket book discount and in transit fare discounts.

Passenger fares for Golden Gate Transit buses and ferryboats remain unchanged for fiscal year 1990-91. Carpool vehicles (with three or more passengers) will continue to pass free across the Bridge during weekday peak commute hours.

## General Manager's Report

**Bridge Division.** (An) engineering firm completed a structural inspection of the Golden Gate Bridge. The project included the inspection of major components of the Bridge and the main cables. Wires from the main cable were tested to establish a baseline for future inspections and analysis of the cable. The inspection found the Bridge to be in good condition.

In line with the Study's recommendation for regular inspection and maintenance of the main cables, the Engineering Department is having movable platforms designed to facilitate painting of the cables.

Radar and Traffic Safety. The Board of Directors voted to purchase a new type of radar that the California Highway Patrol will operate to enforce the 45-miles-per-hour speed limit on the Bridge and its approaches. Older radar units did not perform satisfactorily on the Bridge because of the steel in the structure and the close proximity of other vehicles.

Bridge Traffic. Total vehicle crossings were 42,819,000. Commute traffic was down 2.5 percent, and total traffic was down 2.0 percent.

## Bus Transit Division

On November 9, 1990 Marin County and the City of San Rafael official helped celebrate the groundbreaking for the $3.7 million multi-modal San Rafael Transportation Center on Hetherton Street between Second and Third Streets. The Transportation Center is scheduled to be completed by the end of December 1991. It will serve Golden Gate Transit, Marin Airporter, Whistlestop Wheels, Greyhound Line and taxicabs. The facility was also designed to permit its reconfiguration for future light rail transit.

New Buses. Sixty-three new Transportation Manufacturing Corporation (TMC) suburban coaches are scheduled for delivery in late 1991. The TMC 40-passenger buses, each equipped with one wheelchair position, will replace older GMC buses, which provide 49 seats, but no access for wheelchair users.

The District now operates 256 coaches on regular runs, provides 17 club buses and holds 47 coaches in reserve. There are 360 bus operators employed.

Accessibility. The District continues to expand the accessibility of its bus fleet; the percentage of the fleet which is wheelchair lift-equipped is now about 60 percent, but will reach 84 percent in fiscal year 1991-92.

Bus Patronage Gains. Total ridership was 9,023,742.

## Ferry Transit Division

On August 15, 1990, Sausalito Council Member Robin Sweeny reenacted the ribbon-cutting ceremony that welcomed the first run (20 years earlier) of the District's Sausalito ferry, M.V. Golden Gate. Over the two decades, the vessel has carried almost 16 million passengers and travelled a total of 673,000 nautical miles back and forth to San Francisco.

Ridership. Sausalito Ferry ridership was 496,586; Larkspur ridership was 1,009,303.

## Gift Center

The Golden Gate Bridge Gift Center continues to be a success with gross sales exceeding $1.4 million for the fiscal year. The Gift Center offers over 1,500 collectible and souvenir items. ☐

SIXTY YEARS OF SERVICE

# Highlights from Annual Reports

YEAR: 1991— 1992

## President's Message

The Board and staff put forth tremendous efforts toward implementing the recommendation of the "California Governor's Board of Inquiry on the 1989 Loma Prieta Earthquake" to seismically retrofit the Golden Gate Bridge as one of the Bay Area's transportation structures of regional importance.

Three major activities occurred during FY 1991/92:

The Board accepted the final report (of consultant's study.) The studies developed information necessary to determine the extent of seismic retrofit work required to assure the Bridge can withstand a major earthquake.

The District (worked) to obtain federal grant monies to fund the final design of the seismic retrofit, estimated at $11.5 million. The District has received $5.9 million from the Intermodal Surface Transportation Efficiency Act of 1991 for this purpose. The District will provide the remaining $5.6 million.

The District completed its process of selecting consultant firms to perform the final design work. The firms (selected) are among the most respected in the country. This work is expected to be completed by late-1994.

## Five-Year Fare and Toll Program Phase I

In December 1990, financial projections for District operations through FY 1998/99 indicated that in FY 1991/92 expenditures would exceed revenues by approximately $14 million. One of the primary reasons for the shortfall was the cost of the necessary capital expenditures associated with the seismic retrofit of the Golden Gate Bridge. Another reason was the need to provide subsidies from Bridge tolls for operation of the District's bus and ferry systems. The Golden Gate bus and ferry operations are funded between forty-five and fifty percent by bridge tolls, thirty to thirty-five percent by transit fares, with the remainder being met by federal and state subsidies.

While many Bay Area counties have enacted local sales taxes to support their transportation-related projects and services, Marin and Sonoma counties have not and the Golden Gate Bridge District does not have the authority to levy taxes. Therefore, the use of surplus bridge toll revenue is the only available means to support any financial shortfall in the District's transbay transit activities in the Golden Gate Corridor.

The Board of Directors undertook a comprehensive public and agency review of over twenty bridge toll and transit fare increase alternatives and potential service cut-backs. On May 31, 1991, (they) adopted a Five-Year Toll and Fare Program intended to ensure a balanced budget for all District operations through FY 1995/96. Phase I of the program was implemented on July 1, 1991 when the Bridge cash toll was increased from $2.00 to $3.00.

---

*In the rest of the World . . . .*

## HISTORICAL SNAPSHOT

**Events.** *Riots in Los Angeles following police beating of Rodney King. President Bush announces U.S. will phase out CFC's by 1995. California elects two women U.S. Senators. General Motors announces 1991 loss of $7.2 billion, a world record.*

**Films.** *Howard's End, Malcolm X, The Player, Tous les Matins due Monde.*

**Television.** *In U.S. 60% of homes have cable.*

**Sports.** *Toronto Blue Jays become first team outside U.S. to win World Series. The 25th Olympic Games held in Barcelona.*

**Drama.** *John Guare, Six Degrees of Separation. Tony Kushner, Angels in America.*

**Everyday Life.** *Buckingham Palace announces the separation of the Duke and Duchess of York. In U.S. a Barbie doll that says, "Math class is tough" is pulled from market.*

# 1992

## Phase II

The Board agreed that transit users should pay a fair share to support District transit operations and that transit fares should be increased. However, in order to encourage greater transit usage, the Board deferred for one year, to July 1, 1992 the proposed increase in intercounty bus and ferry fares.

At the close of FY 1991/92, the Five-Year Fare and Toll Program is well underway. As a result, the District is maintaining a balanced budget while continuing the level of operations and services provided by the District.

Also noteworthy is that since implementation of (this) Program, Bridge traffic has decreased to levels formerly experienced in the 1970's.

## General Manager's Report

This was a busy year for the entire District, with numerous activities and successes occurring in the Bridge, Bus and Ferry Divisions. Overall, the District's mark of success was to complete FY 1991/92 with a positive fund balance. Moreover the budget for the next fiscal years has been balanced. This will permit the District to continue to operate the Golden Gate Bridge in a safe and efficient manner as well as all existing bus and ferry services.

## Bridge Traffic

Total vehicle crossings for the year were 41,445,000. Toll revenues totaled $52,764,229, an increase of 36.5 percent over the previous year. The increase was a direct result of the change in the cash toll from $2.00 to $3.00 and the increase in the Bridge toll discount ticket value from $1.67 to $2.22 on July 1, 1991. The additional revenues will provide necessary subsidies for the District's bus and ferry systems, and will provide a portion of the funding needed for the Golden Gate Bridge District seismic retrofit project along with other essential capital projects of the District.

## Bus Division

This was a year of excitement for the Bus Transit Division. Most noteworthy was the 20th Anniversary of the commencement of Golden Gate Transit transbay bus services in January 1992 which coincided with the Grand Opening of the new Transportation Center in San Rafael. After 20 years of using a "temporary" bus stop under Highway 101, the new Transportation Center was constructed in 1991 and opened on January 2, 1992.

The new transit hub provide improved operations with an increase in passenger loading capacity, and more efficient traffic patterns effecting cost savings to the District.

**Bus Ridership.** Bus patronage increased nearly two percent in FY 1991/92. Ridership between Sonoma and Marin counties and San Francisco led the upswing with over four percent growth. Total bus passengers for the year: 9,185,000.

**Replacement Buses.** In 1983, Golden Gate Transit began a program of annually replacing portions of its fleet of 272 buses with new buses. During the fiscal years, 63 new buses, each fitted with passenger lifts, were purchased. Through this program the District is able to maintain an efficient fleet, while increasing the number of buses that are accessible to persons with disabilities as required under State and Federal law. The new buses purchased by the District also have engines which generate less air pollution, thereby helping to achieve clean air standards.

## Ferry Division

**Bay to Breakers Service.** For the first time, the Ferry Division provided service from the Larkspur Terminal to the start of the 1992 Bay to Breakers race in San Francisco. The service was so popular that the Ferry Division plans on providing two vessels for next year's race.

**Ecological Studies.** A four-year study, the "Corte Madera Ecological Reserve Erosion Study," continued this year. The study was undertaken to better monitor the potential impacts of Larkspur vessel on erosion in the Reserve located south of the Larkspur Ferry Terminal.

**Jazz on the Bay.** This year marked the kick-off of the eighth annual "Jazz on the Bay" concerts and the "Lunch for the Office Bunch," two very successful events offered by the Marketing Department for ferry patrons during the summer season. □

SIXTY YEARS OF SERVICE

# Highlights from Annual Reports

YEAR: 1992— 1993

## President's Message

Looking toward FY 1993/94, this District is one of the few public agencies in the San Francisco Bay Area with a balanced budget. This was achieved without changing the Bridge toll structure or significantly impacting service levels. However, arriving at the FY 1993/94 balanced budget required pursuit of a thorough, methodical decision-making process.

Each spring when the Board looks toward the next fiscal year to determine a budget, it looks first at five- and nine-year fiscal projections. During the FY 1993/94 budget process, projections showed a $31.5 million shortfall expected by FY 1997/98. Rather than waiting for the deficit to "hit," the Board made intensive efforts to reduce it by conducting an extensive public review of cost-containing and revenue-enhancing options to bring future budgets into balance.

By the close of FY 1992/93, the Board succeeded in reducing the projected five-year shortfall from $31.5 million to $7.3 million. The Board also arrived at a balanced budget for FY 1993/94 which was $2.5 million less than that of the previous year. Significant "belt-tightening" measures were implemented in all District division.

Now two years into the Five-Year Fare and Toll Program, the seismic retrofit project is on schedule and moving ahead. Final design will be completed in 1994 and construction is slated to last through 1998.

Santa Rosa Facility Dedicated. In August 1992, the District dedicated Golden Gate Transit's Santa Rosa Bus Facility in honor of Director Helen Putnam. Ms. Putnam served on the District Board from 1979 until her death in 1984, and led a life dedicated to community service. She was Mayor of Petaluma, President of the League of California Cities, and a Member of the Sonoma County Board of Supervisors. Ms. Putnam is fondly remembered for her significant contribution to the community and the District. District services remain at the highest levels of efficiency. With thoughtful planning and an eye toward the future, the District will continue to work with the citizens of San Francisco and the North Bay to meet and overcome the challenges that lie ahead.

## General Manager's Message

During FY 1992/93, the District focused on improving efficiency in day-to-day operations, reducing costs, and directing the progress of capital programs to better serve the public. With nearly 1000 dedicated employees, the District effectively maintains one of the landmark structures of our time, the Golden Gate Bridge, and operates one of the nation's foremost transit systems, Golden Gate Bus Transit and Golden Gate Ferry Transit. The District strives to fine-tune its operations and improve communications with its users in order to advance the District's ability to effectively serve the public.

---

*In the rest of the World . . . .*

## HISTORICAL SNAPSHOT

**Events.** *William Jefferson Clinton inaugurated as 42nd President of U.S. Huge fires in Southern California reach suburbs, cause billion dollars in damage. NAFTA treaty approved. Janet Reno becomes first woman Attorney General. Crown Prince Naruhito of Japan marries Masako Owada.*

**Deaths.** *Dizzy Gillespie, Thurgood Marshall, Arthur Ashe, Marian Anderson, Frederico Fellini.*

**Films.** *Groundhog Day, Jurassic Park, Philadelphia, Schindler's List.*

**Everyday Life.** *Buckingham Palace opened to public; fees go for restoration of Windsor Castle. "Hairgate" occurs in Los Angeles when airport runways closed for 40 minutes while President Clinton had his haircut.*

## 1993

In addition to streamlining activities and programs, the District must also focus on the future, looking to meet the growing transportation needs of Bay Area residents and visitors within the resources available. Of primary concern to most North Bay residents is improvement in transportation along the Highway 101 Corridor. The District will continue to be a key player in meeting that challenge.

Quality transportation service is key to the economic, social and cultural vitality of the North Bay. All of the programs and activities advanced this year directly support the District's challenge to continue to improve service.

*[Much of this year's GM Report dealt with acquisition of 139 miles of Northwestern Pacific Rail Road right-of-way, from Novato to Willits in Mendocino County. This decade-long effort, coordinated with the NWP Task Force is aimed at preserving the corridor for public transportation. The rails and right-of-way are owned by Southern Pacific Lines, one of the original foes of the formation of the district in 1928.*

*Another long section detailed the District's continuing efforts to complete all necessary work to achieve 100% compliance with the 1990-mandated Americans with Disabilities Act. The District's Advisory Committee on Accessibility, comprised of community members with a wide range of disabilities, assisted the District in developing and implementing plans.]*

Affirmative Action Program. At the close of the fiscal year, the District employed 873 full-time and 123 part-time, casual and temporary employees. Thirty-six percent of full-time employees are minority, twenty-five percent are female.

## Bridge Division

*For the fiscal year there were 40,860,000 total vehicle crossings.*

**Seismic Upgrade Project.** In January 1993, the final design phase for the seismic retrofit began. The design work is scheduled to be completed by December 1994, at a cost of $11.6 million.

Two contracts were awarded for the final design work. Sverdrup Corporation, Walnut Creek, California, was awarded contract for design of the seismic retrofit of the Fort Point arch, including pylon S1 with tiedown and pylon S2, and the south viaduct include the south anchorage.

T.Y.Lin International/Imbsen & Associates, Inc., a Joint Venture, San Francisco, California, was awarded a contract for design and seismic retrofit

of the suspension bridge, including Marin anchorage, pylon N1 with tiedown, pylon N2, north and south towers and piers, rehabilitation and strengthening of the south pier fender, and wind stabilization of the main span and north viaduct.

Following completion of the final design, the construction phase, estimated at $125 million (1993 dollars), will begin and take approximately 3.5 years to complete. The earthquake retrofit expense is significant, but represents only one-tenth of the replacement cost of the Golden Gate Bridge estimated to be one billion two hundred fifty million dollars (1991 dollars.)

**Railing Replacement.** Approximately 6,557 lineal feet of pedestrian railing, most of which is on the west side, has been deteriorating due to constant weather exposure over the 56 years since the bridge opened. In January 1993 the District awarded the Pedestrian Rail Replacement Contract. The replacement railing is an exact replica of the original and serves to preserve the historical and architectural character of the Bridge. The work began in March and is expected to be completed by November 1993.

**Seismic Monitoring System.** After the Loma Prieta Earthquake of 1989, the Governor's Board of Inquiry recommended seismic instrumentation monitoring systems be placed on various critical structures in the Bay area, and specifically, the Golden Gate Bridge.

The monitoring system will be part of a nationwide seismic reporting system. Following a seismic event, seismologists will be able to analyze date from the monitoring system regarding ground motion and the structural response of the Bridge.

**24-Hour Bicycle Access.** Beginning in November 1992, bicyclists were granted permission to cross the Golden Gate Bridge 24-hours a day.

## Bus Transit Division

During the fiscal year, the District carried 9,087,000 bus passengers.

**East Meets West.** In March 1993, after nearly three years of planning by Golden Gate Transit and the Metropolitan Transportation Commission (and other agencies), Golden Gate Transit began a new regional connection to the East Bay: a three-year "trial" bus service linking San Rafael with the

Richmond and El Cerrito/Del Norte BART stations. Running seven days a week, the Route 40 reached initial performance goals in the first two days of operation. Ridership continued to grow, from 5200 patrons in the first month, to over 8200 patrons after just three months.

**"Whale Bus."** Celebrating and encouraging the use of public transportation, Golden Gate Transit's Marketing Department created the "Whale Bus," a Golden Gate Transit coach with a life-sized portrait of the wayward whale "Humphrey" and his dolphin friends.

Painted by the world-renowned Sausalito environmental artist George Sumner, the Whale Bus encourages the public to share in the benefits of Golden Gate Transit's "Pollution Solution," the spring marketing campaign. Sumner donated his time and paint for the bus, which served on routes throughout the service area. The Whale Bus also surfaced at fairs, festivals and grade schools to promote Golden Gate Transit.

**Capital Improvement Plans.** Looking to further improve San Rafael's C. Paul Bettini Transit Center, the District is exploring the option of purchasing land adjacent to the Transit Center to develop a mixed-use facility that would include parking. By the close of the fiscal year, several sites had been identified, while funding sources continue to be sought.

The newest (bus) replacements will incorporate ADA-approved front-door wheelchair lifts which improve accessibility for persons with disabilities. During the year, the District purchased 41 replacement coaches, all of which are scheduled for delivery in FY 1993/94.

## Ferry Transit Division

*For the fiscal year, Golden Gate ferries carried a total of 1,466,000 passengers.*

**Facility Improvements.** In June 1993, construction was completed at the Larkspur Ferry Terminal to improve vehicle traffic flow exiting the facility. With funding from the City of Larkspur, one lane was added for vehicles exiting westbound, and one lane was improved for vehicles exiting eastbound. As a result, peak commuter hour congestion leaving the Larkspur facility has been reduced from 15 to 20 minutes to only 5 to 10 minutes.

A maintenance dredging project was completed at the Larkspur Ferry Terminal in April 1993. This was the fourth of six dredging events designed to return the Larkspur Channel and turning basin to their original design depth and width. Since the construction of the Larkspur Ferry Terminal in 1976, the District has undertaken dredging on occasions to counter sediments accumulating at a rate of about six inches annually. Upon returning the channel and turning basic to their original design depth and width next fiscal year, it is planned that the U.S. Army Corps of Engineers will assume maintenance dredging responsibility under the Water Resources Act of 1986.

**New Ferry.** The District Board affirmed its commitment to purchase and operate an addition ferry between Larkspur and San Francisco. The new vessel will cut commute time by 15 minutes, making the Larkspur Ferry transbay crossing times comparable to that of person commuting by auto across the Golden Gate Bridge.

**Aloha Friday.** In cooperation with the American Red Cross, a special fund-raising event for the survivors of Hawaii's Hurricane Iniki was held at Larkspur Landing Shopping Center, across from the Larkspur Ferry Terminal. Organized by the Marketing Department, the ferry brought patrons from San Francisco to Larkspur with hula dancing and live music aboard. The highest Friday Larkspur ferry patronage for October 1992 was recorded for the event, and nearly five thousand dollars were raised for the Red Cross.

**Bay to Breakers.** For the second year in a row, additional ferry service during non-commute hours was provided between Larkspur and San Francisco for the annual San Francisco Examiner's "Bay to Breakers" race in May 1993. Because demand was so high for the special ferry last year, two early morning trips were scheduled instead of one. Nearly fifteen hundred patrons used the service. The District Board has authorized this special service on an annual basis.

**S.F. Ferry Building Redevelopment.** The Port of San Francisco is planning to redevelop the historic Ferry Building, and Golden Gate Ferry Division staff has participated in the preliminary design stage. Ferry Division's involvement centers on public access to the Golden Gate San Francisco Ferry Terminal located behind the Ferry Building. Resources Act of 1986. □

## SIXTY YEARS OF SERVICE

# Highlights from Annual Reports

## YEAR: 1993— 1994

### President's Message

During the 1993/94 budget review process, District staff identified a projected shortfall of $7.3 million for the next five-year period. The Board of Directors developed, evaluated and presented eight cost-containment and revenue-enhancing options for public review. After thorough public review the Board adopted three of the eight options.

A modest GGT bus fare increase of 25 cents and 50 cents for the more highly subsidized bus service from Sonoma County was implemented. As GGT ferry fares were significantly below those charged for other services on San Francisco Bay, a modest fare increase of 25 to 75 cents was implemented. Further, the discount offered to frequent transit users was, effective July 1, 1995, reduced from 25 percent to 20 percent. The discount most commonly allowed by other transit providers is 10 percent.

The Board of Directors implemented these nominal fare increases, noting that today GGT customers receive more transit service for their dollar than 15 years ago. If one considers the effect of inflation and applies the Consumer Price Index, in relative terms, transit fares have actually decreased. For example, the average fare for a bus trip from Marin to San Francisco in FY 1979/80 was $1.50. The same trip today is one dollar more which, when expressed in 1980 dollars, is only $1.18 or 32 cents less than the cost in 1980.

Further, in FY 1979/80, the Golden Gate Bridge toll was $1.00. When expressing today's $3.00 toll in 1980 dollars, the effective toll is only $1.42. Effective, this 42-cent increase not only supports the District's regional transit services, but the implementation of important capital projects including the seismic retrofit and mandated improvement under ADA (Americans with Disabilities Act).

*[The final design of the seismic retrofit has been nearly as complex as the structure itself. The complete story of the Retrofit Program may be found at the end of this section of Annual Report Highlights. Meanwhile, the President concludes:]*

Because the Board of Directors focuses on meeting the transportation needs of the future, by making challenging fiscal decisions, with public input, both the world famous Golden Gate Bridge and its associated transit services continue to thrive.

### General Manager's Report

A great deal of the District's progress during the last fiscal year was due to the outstanding contribution of each District employee. Teamwork played a critical role as these dedicated public servants provided the highest quality of service to our customers. The District's accomplishments to date are a tribute to the hard work and professionalism these men and women bring to the job.

*[Most of the District Division report by the GM deals with programs to being about 100% compliance with The Americans with Disabilities Act (ADA), including modifications to facilities and services.]*

---

*In the rest of the World . . . .*

## HISTORICAL SNAPSHOT

**Events.** *Major earthquake hits Northridge area of Los Angeles. Allegations made in Congress regarding Clintons' involvement in Whitewater Development Corporation. Democrats suffer dramatic defeat in mid-term elections.*

**Science.** *Channel tunnel between Britain and France opens.*

**Music.** *Xaver Paul Thomas, Draussen vor der Tur (opera). Mariah Carey, Music Box. REM, Monster.*

**Films.** *Forrest Gump, Four Weddings and a Funeral, Pulp Fiction, Speed.*

**Everyday Life.** *President Clinton who studied at Oxford without taking a degree, returns for honorary doctorate. A truck driver from Maryland commits suicide by crashing a Cessna on the south lawn of the White House.*

### 1994

*Among these was the acquisition of new wheelchair lift-equipped buses, production of a training video for Bus Operators and facilitation of special services to provide "paratransit" assistance between Marin, Sonoma and San Francisco counties. The District's Advisory Committee on Accessibility assisted in successfully developing a wide range of programs and services. Progress continued, too, on the District's continuing efforts with regard to the acquisition of the Northwestern Pacific Railroad right-of-way. The District has taken an active role throughout the acquisition process, but steadfastly maintains that ultimate responsibility for developing and operating the right-of-way for public transportation should be determined by the counties in which it is located.]*

**Affirmative Action Program.** At the close of the fiscal year, the District employed 872 full-time and 116 part-time, casual and temporary employees. Thirty-six percent were minority, 25 percent were female.

**Doyle Drive Reconstruction.** For 1.5 miles, Doyle Drive is a six-lane elevated roadway serving as the southern approach to the Golden Gate Bridge. The District constructed the approach in 1936 on a federal land easement in the Presidio of San Francisco. The land is owned by the U.S. Department of the Army and will soon be transferred to the National Park Service. Today, the roadway is maintained by California Department of Transportation (Caltrans).

Since the early sixties, the District has worked with Caltrans and other agencies to increase safety and decrease congestion on the approach. Over the year, Doyle Drive has deteriorated because of its age, heavy traffic and marine environment. As a result, several local and state agencies are pursuing dialog aimed at rehabilitating the approach.

Part of the current proposal for reconstructing Doyle Drive included lowering the roadway to ground level with two sections tunneling underground. Before a specific engineering design or environmental assessment may be undertaken, a Doyle Drive Intermodal Study was commissioned. The study will help build a consensus about the necessity of rehabilitating Doyle Drive and put a regional perspective on the project. It is expected to be complete by the end of FY 1994/95.

## Bridge Division

*During fiscal yeasr 1993-94, vehicles crossing the Bridge totaled 40,976,000.*

**Toll Plaza Expansion Study.** Over the years, the Bay Area has come to depend upon the Golden Gate Bridge as a vital transportation link between San Francisco and the counties to the north. As a fixed, six-lane roadway, the Bridge cannot be readily expanded to accommodate traffic growth as other highways. To accommodate traffic growth, the District developed a number of nationally recognized and innovative procedures designed to improve the flow of traffic. These include the utilization of reversible lanes, one-way toll collection and free passage for carpools during peak commute hours.

To further the District's goal of managing safe and efficient traffic flow across the Bridge, specifically on weekends, a detailed study to evaluate methods to improve traffic flow by making modifications to the Toll Plaza was initiated. A preliminary report was presented in January 1994.

**East Parking Lot Rehabilitation.** During October 1993, the East Parking Lot, just below the historic Roundhouse Gift Center on the south end of the Bridge, was repaired and refurbished for the safety and convenience of visitors. Due to nearly constant tour bus and visitor traffic in the lot, the asphalt pavement had approached the end of its service life.

**Lead Cleanup.** In 1992, following the identification of undesirably high levels of lead in soil under and adjacent to the Golden Gate Bridge, the District embarked upon the development and implementation of a remedial action plan to remove the elevated lead concentrations.

The Lead Cleanup Program is proceeding in two phases. Phase I focuses on the proposed access, staging and construction areas of the Seismic Retrofit Project. Phase II focuses on potential areas of contamination outside the Phase I area. During the last fiscal year, the District completed Phase I and submitted a Draft Remedial Action Plan to (the California Environmental Protection Agency).

**Golden Gate Bridge Earns Seven Wonders Designation.** Nearly 57 years after the design and construction of the Golden Gate Bridge, the world-renowned structure continues to garner awards as an American Landmark. In recognition and honor of the nation's "most spectacular civil engineering achievements," the American Society of Civil Engineers designated the Golden Gate Bridge as one of the "Seven Wonders of the United States" in conjunction with National Engineers Week, February 20-26, 1994.

The Golden Gate Bridge shares this prestigious distinction with Hoover Dam, Interstate Highway System, Kennedy Space Center, Panama Canal, Trans-Alaska Pipeline and World Trade Center.

The criteria used in determining the Seven Wonders include "service to the well-being of people and communities, uniqueness, pioneering aspects in design and construction, extent to which the work has become a benchmark for later projects, and great size and beauty." Like the seven ancient wonders, the Seven Wonders of the United States are all works of civil engineers.

**Distinguished Building Award.** In October 1993, in recognition of "enduring excellence in design," the Society of America Registered Architects (SARA) honored the Golden Gate Bridge with its "Distinguished Building Award." For the first time in SARA's history, the Distinguished Building Award was presented to a structure other than a building.

The Golden Gate Bridge was honored because of its "impact on the city, design, economic value, cultural statement, engineering accomplishment and contribution to the overall furtherance of the region." The award commended the work of Bridge Architect Irving Morrow, Chief Engineer Joseph Strauss and the Golden Gate Bridge, Highway and Transportation District.

## Bus Transit Division

*During the fiscal year, Golden Gate Transit buses carried a total of 8,969,000 passengers.*

**"Whale Bus" Awards.** Golden Gate Transit received two awards for Marketing and Public Information department programs associated with the "Whale Bus," a GGT coached painted with a life-size portrait of "Humphrey the Humpback Whale."

In October 1993, the American Public Transit Association named GGT to its 1993 Chairman's Silver Honor Roll for efforts in educating the public about the benefits of public transit and organizing support within the community for improved transit service.

(That same month) the Metropolitan Transportation Commission presented an Award of Merit acknowledging the "Whale Bus" and "Pollution Solution" campaign for significantly contributing to the improvement of transportation.

## Ferry Transit Division

*In fiscal year 1993-94, Golden Gate Transit ferries carried a total of 1,406,00 passengers.*

**Vessel Maintenance.** Each of the four Golden Gate ferries is the focus of a rigorous maintenance program to assure reliable and safe mechanical operation. In addition to daily and weekly inspections, each spring two ferries enter dry dock for U.S. Coast Guard inspection and routine hull maintenance and repairs. During June and July, the Larkspur vessel M.S. Sonoma and the Sausalito vessel M.V. Golden Gate entered dry dock one at a time without significant impact on commuters. While each vessel was out of service, GGT buses provided alternate service.

**New Ferry.** Plans are moving ahead to acquire and operate a new ferry between Larkspur and San Francisco. The new vessel would expand the Larkspur fleet from three to four. Two reports, an analysis of a new vessel's potential impact on the environment and a study of possible access improvements to the Larkspur Ferry Terminal are currently underway.

**Fourth Berth, Larkspur.** To accommodate the overnight storage, service and revenue operation of a fourth vessel at the Larkspur Ferry Terminal, an additional berth was constructed during the fiscal year. Phase I of the project involved dredging an appropriate area in the lagoon between the ferry terminal and East Sir Francis Drake Boulevard. Phase II included the fabrication and installation of a floating steel dock and walkway just west of Berth Three. The project was completed in June 1994. □

SIXTY YEARS OF SERVICE

# Highlights from Annual Reports

YEAR: 1994— 1995

## President's Message

The Golden Gate Bridge, Highway and Transportation District operates a transportation network in the U.S. Highway 101 corridor comprised of the Golden Gate Bridge and Golden Gate Bus and Ferry Transit. The District continues, with limited resources, to provide effective mobility through this network. Two key projects advanced this year, one aimed at sustaining and improving the transportation infrastructure, the other at expanding the transportation network.

**Seismic Retrofit.** In 1990, the California Governor's Board of Inquiry on the Loma Prieta Earthquake called for all important transportation structures in the State to be seismically retrofit to assure continued operation following a major earthquake. The District, in concert with seismic engineers, had already launched a state-of-the-art retrofit design of the Golden Gate Bridge scheduled to be complete by the end of 1995. The Bridge will be retrofit to withstand a Richter magnitude 8.3 earthquake and be functional for emergency vehicles within only a few hours.

The Federal Highway Administration financed $5.9 million of the $13.3 million retrofit design costs. The District's Five-Year Fare and Toll Program generated the remaining $7.4 million. The Program will also generate approximately $30.1 million for seismic construction. That, and $4.9 million from a petroleum anti-trust settlement earmarked for seismic retrofit projects in California, will provide the 20 percent local match required for federal grants being sought to fund the $175 million construction total.

The District is working to secure the remaining 80 percent through federal funds which, as part of the seismic construction effort, will contribute to economic growth in the region by creating an estimated 3,100 jobs. Moreover, a federal investment of $140 million will help preserve an existing public investment in the Golden Gate Bridge, now estimated to have a replacement value of more than $1.4 billion.

**Northwestern Pacific Railroad Right-of-Way.** The District's 1970 long-range transportation plan called for the preservation of the NWP right-of-way as a future corridor for public transportation. In 1990, the District joined with Marin and Sonoma counties to acquire 152 miles of right-of-way from Southern Pacific, extending from Route 37 in Novato north to Willits and east from Novato to Lombard. In 1987 and 1993, Congress authorized funds for 80 percent of the $27 million purchase price. The remaining 20 percent will come from state funds.

During the past year, the District, Marin County and North Coast Railroad Authority formed the Northwestern Pacific Railroad Authority to take title of the right-of-way in 1996. The counties in which the right-of-way is located will set policy for developing public transportation along (it).

## General Manager's Report

As the effects of public service belt-tightening rippled across the country, the District continued to fulfill its public trust with increasingly limited resources. Despite this challenging fiscal environment, many important projects moved ahead this year.

**New Buses.** Bus Transit received 41 new Flexible coaches. The purchase is part of an ongoing bus replacement program retiring older buses and replacing them with new, "clear-air" fuel-efficient models that comply with Americans With Disabilities Act (ADA) regulations. Golden Gate Transit's bus fleet is now 98 percent lift-equipped and scheduled to be 100 percent lift-equipped by FY 1996/97.

## 1995

*During the year Golden Gate Transit buses carried a total of 8,733,000 passengers.*

**Full Service Paratransit.** ADA regulations require the District to provide special intercounty van service for persons with disabilities who cannot use fixed-route buses. The District became the first public transit operator in the Bay Area to fully implement its paratransit program and attain full compliance with ADA.

**New Larkspur Ferry.** The District proceeded with environmental impact studies for the addition of a fourth vessel to the Larkspur fleet. A high-speed ferry is proposed to reduce the Larkspur to San Francisco crossing time from 45 to as little as 30 minutes and to increase daily crossings. The expansion is part of the Metropolitan Transportation Commission's 1991 Regional Ferry Plan and is anticipated to be funded by federal and state grants.

*During the year, Golden Gate ferries carried a total of 1,332,000 passengers.*

**Toll Plaza Expansion.** Since the Bridge's six-lane roadway cannot easily be expanded like other highways, in FY 1993/94 the District began studying potential Toll Plaza modifications to improve traffic flow, especially on weekends. The study has resulted in two scheduled improvement. First, beginning in the spring 1996, the Toll Plaza roadway is to be repaved. Second, beginning spring 1997, the roadway east of the Toll Plaza is scheduled to be widened and traffic realigned to accommodate an additional toll booth.

*Total crossings of the Golden Gate Bridge this year totaled 40,715,000.*

Contracts totalling $15.2 million were awarded for professional services, construction, supplies, equipment and other services, with 14 percent awarded to Disadvantaged Business Enterprises (DBE). Of the $3.8 million in contracts which received funding through the U.S. Department of Transportation, 29 percent were awarded to DBEs.

To optimize public investment, efficiency has always been the rule by which District decisions are measured: every purchase, every bus and ferry, every service.

Every hour of every day, District services are there. From the Route 80 leaving Santa Rosa at 4 a.m., to the popular 5:20 p.m. Larkspur Ferry, or Bridge tow operations during the graveyard shift, District services must be reliable.

Our transportation network is a lifeline to jobs, health care and other services. Reliable mobility is a measure of the quality of life and the competitiveness of our economy.

Golden Gate Transit has one of the best reliability records nationwide. Combined, the transit system operated 99.9 percent of scheduled service with more than 92 percent on- time reliability.

Our buses average more than 28,500 miles between mechanical failures and Golden Gate ferries run more than 15,400 miles between breakdowns. The latest national average for buses is less than 4,800 miles between failures. □

SIXTY YEARS OF SERVICE

# Highlights from Annual Reports

YEAR: 1995— 1996

[As we near the present, here once again is the mission statement of the District. From time to time, it has been included in the Annual Reports with minor changes to reflect mandated changes by new State and Federal laws. This is the core statement by which the Board of Directors continues to direct the activities of the Management and Staff of the District.]

## Mission Statement

Since 1937, the Golden Gate Bridge, Highway and Transportation District has served the public interest by operating and maintaining the world famous Golden Gate Bridge across the entrance to San Francisco Bay. As part of U.S. Highway 101, the Bridge serves as a vital transportation link between the City of San Francisco and the vast Redwood Empire to the north.

In 1969, The California State Legislature expanded the District's transportation responsibilities to include the operation of safe, efficient, and cost-effective public transit in the Golden Gate Corridor north of San Francisco. Since 1970, Golden Gate Bus and Ferry Transit has been a viable alternative to the automobile, while contributing to the protection of our environment.

The District is unique among Bay Area transit operators because it provides transit service without support from local sales taxes or dedicated general funds. Since the District has no authority to levy taxes, surplus Bridge toll revenue is the only local source of funding for transbay transit service. Currently, Golden Gate Bus and Ferry Transit is funded 50 percent by Bridge Tolls and 30 percent by fares. Federal, state and local subsidies, advertising, and property rental revenues meet the remaining costs.

Based in San Francisco, the District consists of three operating divisions, Bridge, Bus, and Ferry, and an administrative District Division. Overseeing 917 employees (37 percent minority and 24 percent female) working together in the public interest, the General Manager coordinates the operations of all divisions according to the policy and direction of the District Board of Directors. The board consists of nineteen members representing six counties: San Francisco, Marin, Sonoma, Del Norte, and parts of Napa and Mendocino.

The District's mission is to provide safe, efficient, and reliable means for the movement of people, goods, and services within the Golden Gate Corridor. In carrying out this mission, the District operates and maintains the Golden Gate Bridge in a structurally sound condition to provide safe and efficient travel for vehicles and other modes of transportation; provides public transit service, such as buses and ferries, which operate in a safe, affordable, timely and efficient manner; and carries out its activities in a cost-effective, fiscally responsible manner. The District recognizes its responsibility to work as a partner with federal, state, regional, and local governments and agencies to best meet the transportation needs of the people, communities, and businesses of San Francisco and the North Bay Area.

The District's goals are to maintain the Golden Gate Bridge; ensure reasonable mobility across the Bridge by providing for public transit service and encouraging Ridesharing as far as resources permit; and contribute to the protection of the environment by working as a partner with other public agencies in providing attractive, efficient regional public transit services as an alternative to the private automobile and by encouraging the use of such services.

*Total vehicles crossing the Golden Gate Bridge for the fiscal year 1995/96: 41,267,000. Golden Gate Transit buses carried 8,945,000 passengers. Ferry passengers during the year, 1,432,000.*

## 1996

## President's Message

**NWP Right-of-Way Purchase.** It took two decades of steadfast commitment and effort by numerous local, state, and federal policy makers to make it happen. In April 1996, the Board of Directors of Northwestern Pacific Railroad Authority adopted the final resolutions needed to complete purchase of the 139 mile right-of-way.

The terms of the purchase agreement were signed on April 11, 1996 and the transaction was completed on April 29, 1996. The purchase was deemed a truly historic moment for the future of transportation in the Redwood Empire as the purchase preserves the right-of-way for generations to come.

**Seismic and Wind Retrofit Project.** The Golden Gate Bridge stands at the entrance to the San Francisco Bay as an international treasure and vital link in California's highway system. Since opening in 1937, it has served as a catalyst in promoting the economic vitality of the San Francisco Bay Area. A natural disaster such as a major earthquake could cause severe, perhaps irreparable damage to the historic structure, dealing a debilitating blow to the region's economy and quality of life.

What took four years to build could, in only minutes, be destroyed.

The 1989 magnitude 7.1 Loma Prieta Earthquake occurred 60 miles to the south, causing severe damage to many significant structures in the Bay Area. The Bridge suffered no damage from the moderate and distant quake. But the wake-up call had come. Redesigning the Bridge to withstand a maximum credible earthquake of 8.3 magnitude occurring on the nearby San Andreas Fault is imperative.

*[Note: A complete report on the Seismic Retrofit Program follows these fiscal year highlights.]*

## General Manager's Report

Meeting the challenge of efficiently and safely operating a complex transportation network serving more than 60 million customers annually requires a steadfast focus and careful management of the District's fiscal resources. By investing a portion of these resources in the preservation of our transportation infrastructure, the District aids in sustaining the region's economic vitality.

## Infrastructure Improvements

**New Buses Ordered.** In February 1996, the Bus Transit Division began the process for purchasing 30 newly-designed Motor Coach Industries (MCI) "Commuter Special" extra-long buses. The MCIs (are) "clean-air" buses that accommodate persons in wheelchairs. AT 45-feet long, the new MCI is five feet longer than the GGT's standard bus. With their greater seating capacity (they) will provide greater operating efficiency. The extra length allows for a total of 57 passenger seats, topping the current fleet average by 16 seats.

**Sausalito Ferry Terminal.** In March 1996, the District began renovating the Sausalito Ferry waterside landing located at Humboldt and Anchor streets in Sausalito. Two significant improvements will be achieved once the project is completed in September 1996. First, a new pedestrian gangway designed to accommodate wheelchair users and meet ADA regulations will replacing the ageing gangway. Second, a new wider floating dock allowing two ferry vessels to dock at the same time will be constructed.

## Other Programs & Activities

**Ferry Anniversary.** On August 20, 1995, the Golden Gate Ferry Division celebrated 25 years of service from Sausalito to San Francisco. For the past quarter century, Golden Gate Ferry has made a lasting contribution to the region by helping to reduce traffic on the Golden Gate Bridge and promote interests that are important to us all: air quality, conservation and protection of the environment, reduced energy consumption, and safe, efficient cost-effective transportation.

Through the years, Golden Gate Ferry has fulfilled a public need through bad times and good. From the Marin County floods of '82 to the earthquake of '89, Golden Gate Ferry has played an important role in maintaining regional transportation. From the summertime "Lunch Bunch" and "Jazz on the Bay" series to the holiday "Merry Ferry," Golden Gate Ferry has become an integral part of life in Marin and San Francisco counties. □

# GOLDEN GATE BRIDGE
# SEISMIC RETROFIT DESIGN

## Today's Technology Transforming Yesterday's Engineering

The Golden Gate Bridge is internationally renowned, recognized as one of the seven engineering wonders of the U.S., and distinguished as one of the greatest suspension spans ever built. Because of its excellent design, its history of significant structural improvements and ongoing maintenance program, the Bridge has a life span estimated at 200 years. However, should a major earthquake (magnitude 7 or greater) with a nearby epicenter occur, the Bridge could fail. The purpose of the Golden Gate Bridge, Highway and Transportation District seismic retrofit project is to apply modern engineering knowledge, methods and materials to improve the Bridge's ability to withstand damage from strong nearby earthquakes.

## Getting Started.

In the aftermath of the 1989 Loma Prieta Earthquake, the District contracted with T.Y.Lin International to undertake a seismic vulnerability study of the Golden Gate Bridge. The Bridge suffered no damage from the 1989 distant earthquake. As the initial studies commenced, a Board of Inquiry on the Loma Prieta Earthquake, convened by the Governor of California, issued its 1990 report directing major transportation structures be retrofitted to assure their functionality following future major earthquakes.

Engineers first reviewed the original design principles applied to the Bridge and structural improvements made since its opening. Joseph B. Strauss designed the Golden Gate Bridge to move in order to disburse and absorb effects from traffic, weather, and ground movement. It expands with sunshine and contracts with cold or rain and moves up or down with varying traffic loads. Even though the Bridge was built using the most advanced theories of suspension bridge behavior of the time, a review of the original design for seismic forces determined that improvement is needed. Over its 59-year life, the Bridge has undergone several structural improvements. Most noteworthy were the addition of a bracing system under the roadway, the replacement of the suspender ropes, and the replacement of the roadway deck.

Having studied the Bridge's original design and structural capabilities, the next step was to conduct a vulnerability study. First, engineers did ground motion studies. They analyzed the geology and topography adjacent to the Bridge. They conducted seismic risk investigations examining earthquake probability, strength, and proximity. Ultimately, varying earthquake conditions were modeled on computers to analyze the Bridge's behavior.

The vulnerability study showed that although the Golden Gate Bridge has performed well in all past earthquakes, "It is vulnerable to damage in a Richter magnitude 7 or greater earthquake with an epicenter near the Bridge, and it could be closed for some time after such an earthquake." Engineers determined that both the south and north approach viaducts were among the most vulnerable to collapse under a strong nearby earthquake. Preliminary analysis showed that in a strong nearby earthquake the tower saddles, which support the main cables where they cross over the tower tops, would be subject to movement and possible damage. Further, the suspension span could ram the towers, the Fort Point arch could uplift and move from its current location and the lightly reinforced south pylons would be subject to extensive damage.

## Establishing Performance, Architectural, and Design Criteria

The next stop for engineers was to evaluate Bridge structures against performance criteria specifically developed for the Golden Gate Bridge. Bridge structures meeting the criteria would not require retrofitting, while those that did not, would. The performance criteria included not only consideration of engineering issues, but public policy, social and economic considerations, including the historical, cultural and architectural values of the Bridge and its role as a vital economic force in the Bay Area.

The performance criteria were applied to the Golden Gate Bridge on that basis that it must:

- 1. Withstand a maximum credible earthquake equivalent to a magnitude 8.3 earthquake occurring on the nearby San Andreas Fault (7 miles west of the Golden Gate Bridge).

- 2. Permit access for emergency vehicles.
- 3. Be available for limited vehicular access within a few days after the earthquake.
- 4. Be fully operational within one month.

Further, maintaining traffic flow during retrofit construction and maintaining the Bridge's aesthetic and historic integrity were essential to the retrofit's success. Requirements were also developed to assure preservation of the historic architectural value of the Bridge.

Since engineering design criteria did not previously exist for seismic retrofit of long-span suspension bridges, design standards we developed based on state-of-the-art seismic engineering principles, the performance criteria, and architectural requirements. Questions assessed included, "What level of damage would be acceptable for meeting the performance criteria?" and "What technology should be used to assess the Bridge's seismic performance?" Design criteria allowed engineering analysis to relate the known behavior of the Bridge with behavior determined by physical testing or modeling. Further, the criteria allowed for identification of the limits of repairable damage that would not threaten structural safety and could be repaired without interrupting traffic.

## Developing Retrofit Methods

With initial seismic vulnerability studies completed, performance criteria determined, architectural requirements established, and design standards set, engineers next had to decide how to best retrofit the Golden Gate Bridge. To accelerate

this final design effort, two leading engineering firms, T.Y.Lin International/Imbsen & Associates, Inc., a Joint Venture, and Sverdrup Corporation, worked concurrently on different components of the Bridge.

Engineers conducted sophisticated computer modeling analysis to identify in much greater detail how the different structures of the Bridge would behave under the violent forces of a major nearby earthquake. They then identified means to control the behavior. Engineers developed ways to minimize the impact of seismic energy by giving the Bridge structures more flexibility to move and dissipate forces. At the same time, the structure had to be strengthened to bear colossal forces. Various retrofit schemes were modeled and the best alternatives identified.

Computer analysis determined the Golden Gate Bridge's performance could be enhanced by tuning elements of the structure to reduce violent action caused by ground motions of a nearby earthquake and by strengthening certain of the structures to reduced the damage caused by these actions. It was demonstrated that resisting the forces by strengthening would be extremely expensive, less effective and would compromise the aesthetic integrity of the Bridge. The tuning of structures to move with the seismic forces, rather than resist them, dramatically reduced the amount of strengthening required and the overall cost of the retrofit. In addition, engineers conducted extensive physical component testing on scale models to confirm performance of several of the proposed retrofit measures.

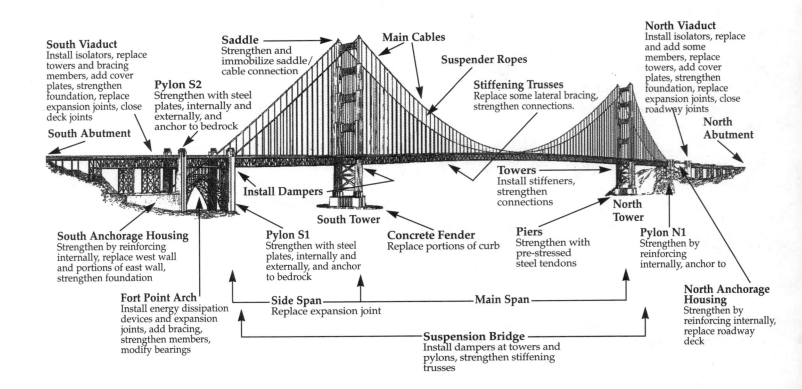

**South Viaduct**
Install isolators, replace towers and bracing members, add cover plates, strengthen foundation, replace expansion joints, close deck joints

**Pylon S2**
Strengthen with steel plates, internally and externally, and anchor to bedrock

**South Abutment**

**Saddle**
Strengthen and immobilize saddle/cable connection

**Main Cables**

**Suspender Ropes**

**Stiffening Trusses**
Replace some lateral bracing, strengthen connections.

**North Viaduct**
Install isolators, replace and add some members, replace towers, add cover plates, strengthen foundation, replace expansion joints, close roadway joints

**North Abutment**

**Install Dampers**

**South Tower**

**South Anchorage Housing**
Strengthen by reinforcing internally, replace west wall and portions of east wall, strengthen foundation

**Pylon S1**
Strengthen with steel plates, internally and externally, and anchor to bedrock

**Concrete Fender**
Replace portions of curb

**Towers**
Install stiffeners, strengthen connections

**Piers**
Strengthen with pre-stressed steel tendons

**North Tower**

**Pylon N1**
Strengthen by reinforcing internally, anchor to

**Fort Point Arch**
Install energy dissipation devices and expansion joints, add bracing, strengthen members, modify bearings

**Side Span**
Replace expansion joint

**Main Span**

**Suspension Bridge**
Install dampers at towers and pylons, strengthen stiffening trusses

**North Anchorage Housing**
Strengthen by reinforcing internally, replace roadway deck

## North Viaduct Present Condition

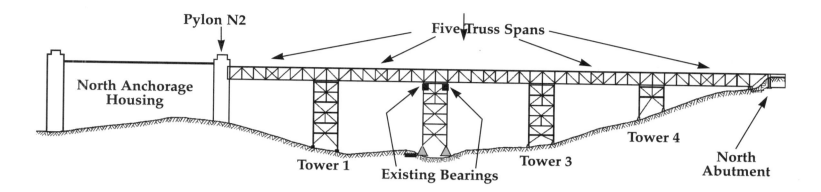

Pylon N2

Five Truss Spans

North Anchorage Housing

Tower 1

Existing Bearings

Tower 3

Tower 4

North Abutment

## North Viaduct Retrofit Scheme

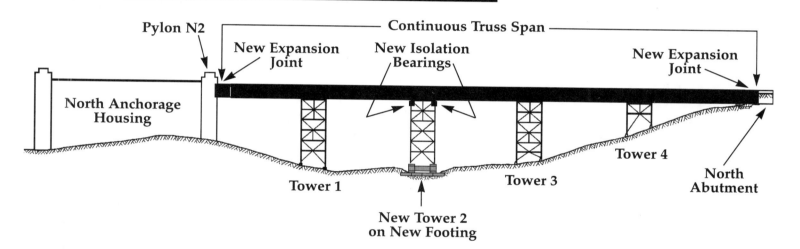

Pylon N2

Continuous Truss Span

New Expansion Joint

New Isolation Bearings

New Expansion Joint

North Anchorage Housing

Tower 1

New Tower 2 on New Footing

Tower 3

Tower 4

North Abutment

## South and North Viaduct

*Retrofit measures proposed for the south viaduct are similar to those proposed for the north viaduct. The following is a more detailed look at measures proposed for the north viaduct. The north viaduct extends from the north anchorage housing to the north abutment. It consists of five independent truss spans supported by four steel towers, pylon N2 on the sound end, and the north abutment at the north end. At the top of each of the support towers, the spans are supported by a fixed steel bearing on one end and a moveable rocker bearing at the other. Each of the five truss spans is subject to failure from the movement of the support towers, which would cause failure of the bearings if seismic forces lift the spans off their bearings.*

*The five truss spans will be tuned by joining them to form a continuous span. This ties the truss system together so it can move as one unit and better distribute seismic forces at the support towers. The steel members of the spans will also be strengthened. At either end of the now continuous span, new expansion joints will be added to allow the span to move relative to the north abutment and pylon N2 to prevent transfer of damaging forces. Isolation bearings, consisting of laminated neoprene layers with head cores, will replace the existing bearings at the top of the new, stronger support towers on new, stronger footings and foundations. The isolation bearings will dramatically reduce the transfer of seismic forces applied to the entire viaduct structure and therefore reduce the amount of strengthening required throughout the structure.*

## Fort Point Arch and Pylons S1 and S2

The Fort Point arch spans 320 feet over historic Fort Point, the only example of an American Civil War fortification on the western coast of the U.S. Framing the arch are two 220 foot tall art deco pylons of lightly reinforced concrete cellular construction. The pylons provide hold down for the main cables of the suspension bridge to direct their passage through the arch to the south anchorage.

Fort Point arch is supported on rocker bearings at its base and is subject to uplift during a major earthquake. Analysis determined the arch should be tuned to be allowed to uplift and then be redirected back onto its bearings. This approach greatly reduced the need to strengthen the arch. The Fort Point arch pylons will also be retrofitted to provide for controlled rocking in a seismic event, permitting the dissipation of seismic energy. The pylons will be strengthened, internally and externally, with steel plates. To preserve their original appearance, the external concrete surface will be restored. The pylon foundations will be tied to bedrock level.

Unique to the retrofit scheme is the installation of energy dissipation devices (EDDs) to dissipate seismic forces on Fort Point Arch. During an earthquake, EDDs would provide a point where energy would be dissipated, thereby avoiding potential failure of the arch structure itself. With this design the entire arch is allowed to move, but the movement is controlled. The design calls for EDDs between the top of the arch and the pylons to dissipate side-to-side motion while controlling the extent of uplift. Other EDDs are placed between the pylons and the arch to dissipate length-wise seismic energy. The EDDs consist of plates of dissimilar materials. They dissipate energy by means of friction between the sliding surfaces.

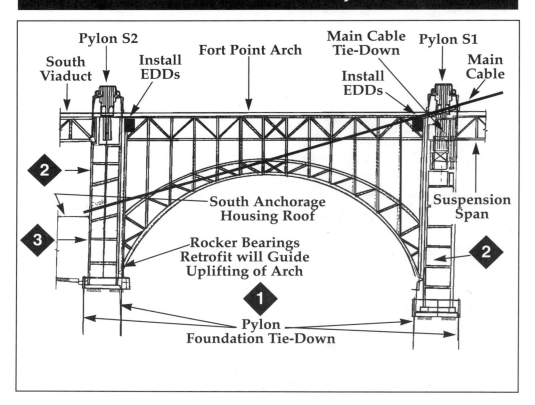

**Retrofit of the Fort Point Arch and Pylons S1 and S2**

**1** Tie pylon foundations to bedrock level.

**2** Install steel plating on exterior and interior wall faces of pylons. Provide cross tie anchor bolts through existing concrete walls.

**3** Provide isolation joint between south anchorage housing and pylon S2.

## Suspension Bridge and Towers

The 6,450 foot suspension bridge consists of the main span and two side spans. The stiffening trusses of the spans are suspended from the towers by the main cables and suspender ropes. Reinforced concrete pylons support the stiffening trusses at the ends. The suspended structure consists of stiffening trusses connected by both a top and bottom side-to-side bracing system. The suspended structure is connected to the towers and pylons through wind-locks. The wind-locks are designed to control the movement of the suspension span caused by the wind. The main cables are supported on the towers in cast steel saddles. The 746-foot-tall towers are of multi-cellular construction with riveted plates and angles.

Analysis determined that fluid viscose dampers installed horizontally between the suspension bridge truss and the towers would prevent the suspension span from slamming into the towers under a severe earthquake. Dampers would also reduce stresses on the towers and suspension spans, and isolate the pylons from seismic forces. During design analysis, engineers recognized that there was little precedent for using fluid viscous dampers in the seismic retrofit of bridges. Scale models were built and tested to verify the performance of the dampers. In addition, the top

lateral bracing systems of the suspension span will be strengthened to prevent buckling of steel members.

Computer modeling and physical tests performed on the scale models showed the tower legs would rock on their base in a severe earthquake. A smaller part of the tower leg was then analyzed, revealing the interior cells of the tower leg would be subject to buckling. As a result, critical locations of the tower legs will be stiffened to prevent buckling. The retrofit calls for the towers to be allowed to still lift off their bases, but the legs are strengthened so they will not crumple. The concrete piers supporting the towers will also be reinforced.

Once in place, the newly advanced seismic retrofit measures will transform the Golden Gate Bridge, built before the advent of modern seismic engineering, into a new modern structure. Carrying over 40 million vehicles a year, the seismically retrofitted Golden Gate Bridge will lead the way for other projects around the world, applying state-of-the-art techniques surpassing any suspension span retrofit efforts to date. □

# THE
# BRIDGE
## A CELEBRATION

# LATE NEWS & UPDATES.

## New Ferry Ordered

**November 8, 1996.** The District awarded a contract for a new ferry to a Washington state firm. The new addition, to be named *M.V. (Motor Vessel) Del Norte,* is scheduled to be placed in service in the Golden Gate Larkspur Ferry Fleet by **January 1988.**

The 325-passenger boat is a fast, light-weight catamaran designed by Advanced Multihull Designs Pty. Ltd. of Sydney, Australia. It will feature two decks, with seating for 92 passengers on the upper deck, plus room for four wheelchairs. There is seating for 212 on the lower deck and 21 outside.

The ferry is 135 feet long, approximately 35 feet shorter than the current Larkspur vessels. It is 20 feet longer than the Sausalito Ferry. It will cruise at 35 knots. Larkspur ferries cruise at 20.5 knots while observing strict speed limits in the Larkspur Channel, it will nevertheless reduce the commute between Larkspur and San Francisco from 45 to 30 minutes.

Anticipating increased ferry use, during the Summer of 1997, the District will add 157 parking spaces at the Larkspur Terminal.

## Transit District Celebration

**January 1997.** Golden Gate Transit (GGT) celebrated its 25th Anniversary. Now serving more than nine million passengers annually, GGT has become an integral part of daily life in the North Bay. Because of the development of GGT, traffic growth in the Golden Gate Corridor has been held to a manageable level. Prior to the start of GGT, 30,000 people in 20,000 vehicles crossed the Golden Gate Bridge during each morning commute. By the end of 1996, over 35,000 people were crossing the Bridge each morning, but vehicle numbers had grown to only 20,900. GGT has continued to fulfill the mission of reducing automobile traffic and congestion while contributing to the protection of the environment with efficient, reliable and cost-effective alternatives to the private automobile.

## A Terminal Move

**March 2, 1997.** Golden Gate Transit moved its San Francisco Civic Center/Transbay Terminal operations to Mission Street to improve operating efficiency and for the convenience of patrons closer to downtown destinations. The move also enhances opportunities for passengers transferring to BART, SamTrans and MUNI.

## New Wheels

**March 3, 1997.** GGT placed 30 newly designed Motor Coach Industries (MCI) "Commuter Special" coaches into service on heavily patronized commute routes to increase operating efficiency. The 45-foot long coaches are five feet longer than GGT's standard bus. The coaches accommodate 57 passengers — 16 more than the fleet average. There are two wheelchair positions on each coach. The purchase was funded through federal and state grants.

## Toll Plaza Improvements

**March 21, 1997.** Repaving of the Golden Gate Bridge Toll Plaza is completed. In order to maintain the flow of traffic during peak weekday commute periods, and on Friday nights, all work was performed at night.

## Lane Barrier Studies

**March 26/27.** A new, one-foot median barrier prototype, developed by Barrier Systems, Inc., undergoes independent certification testing. The system is being evaluated for possible use on the Bridge. In **July 1996,** the District agreed to contribute up to $42,500 to the cost of testing. Northwestern University Traffic Institute is performing a comparative risk assessment of the new one-foot barrier to determine its potential benefit of preventing lane crossover accidents against its potential for increasing other types of accidents or reducing traffic capacity.

**April 8, 1997.** Ten competitive bids with favorable pricing were received for the first of several phases of seismic retrofit construction. Construction is scheduled to begin this summer. The two and one-half year first phase will retrofit the Marin approach and clean up historic lead contamination in soil in construction areas. Bridge traffic and the Bridge's aesthetic and historic appearance will be unaffected by the retrofit construction.

## Retrofit Project

*[The following is from the May/June 1997 issue of "Golden Gate Gazette," a publication produced by the Golden Gate Bridge, Highway and Transportation District for its customers.]*

### Toll Dollars Fund First Phase of Retrofit Construction

Retrofitting the Golden Gate Bridge to withstand a maximum credible earthquake of 8.3 occurring on the nearby San Andreas and Hayward Faults is imperative. Designed before the advent of modern seismic engineering, the Bridge serves as a catalyst in the economic vitality of the Bay Area. Carrying over 40 million vehicles per year, it is estimated that over nine million people from around the world visit the Bridge each year. Bridge failure would cause severe, unrecoverable damage to the historic structure, dealing a debilitating blow to the region's economy and quality of life.

The $13.3 million, four-year seismic retrofit design was completed in **January 1997.** The first of four construction phases is planned to start this summer. While federal funding necessary to complete the project has not yet been secured, the $30 million first construction phase will be funded primarily through toll revenues set aside since 1991 when the $3.00 toll was instituted. In addition, the District, in exploring other funding sources, intervened in an anti-trust petroleum litigation whose settlement funds were earmarked for seismic retrofit projects in California. The District secured $4.9 million for Bridge retrofit. The first phase of construction will include clean up of historic lead in construction areas and Marin approach retrofit work.

The District continues to seek federal funds totaling $140 million to complete the remaining phases. Once funding is secured, the remaining construction work could be completed in three to five years. It is estimated that the retrofit will create 3,100 jobs over the construction period. Moreover, the requested federal investment of 80 percent of the estimated $175 million total construction cost will help preserve the existing public investment in the Golden Gate Bridge, now estimated to have a replacement value of $1.4 Billion (1997 dollars).

**May 1997.** Continuing a tradition, Golden Gate Ferry operates late night service between San Francisco and Larkspur, serving the bi-annual Black and White Ball benefit. Golden Gate ferry service also aided another longtime San Francisco benefit project, the annual Bay to Breakers Race, by adding special early-morning service between Larkspur and San Francisco. **May 27.** The Golden Gate Bridge observed its 60th Anniversary.

**October 1997.** Golden Gate Transit is scheduled to receive 30 new 40-foot suburban 43-passenger coaches from Nova Bus. Funding for these coaches is through federal and state grants.

## Continuing Commitment

"Transportation is an integral part in each of our lives. The District is proud to be a partner in serving the transportation needs of the Bay Area through operation of the Bridge, buses and ferries. We will continue working closely with the counties and cities as plans emerge for the development of the former Northwestern Pacific Railroad right-of-way running from Healdsburg to Larkspur, California."

Carney J. Campion,
General Manager
Golden Gate Bridge, Highway
and Transportation District.